Statistics and Computing

Springer Science+Business Media, LLC

Statistics and Computing

Yvan Pannatier

VARIOWIN

Software for Spatial Data Analysis in 2D

With 37 Illustrations

 Springer

Yvan Pannatier
Institute of Mineralogy
University of Lausanne
BFSH2
1015 Lausanne, Switzerland

Series Editors:

J. Chambers
AT&T Bell Laboratories
Murray Hill, NJ 07974
USA

W. Eddy
Department of Statistics
Carnegie Mellon University
Pittsburgh, PA 15213
USA

W. Härdle
Institut für Statistik und
　Ökonometrie
Humboldt-Universität zu Berlin
D-10178 Berlin, Germany

S. Sheather
Australian Graduate School
　of Management
Kensington, New South Wales 2033
Australia

L. Tierney
School of Statistics
University of Minnesota
Minneapolis, MN 55455
USA

Printed on acid-free paper.

Production managed by Laura Carlson; manufacturing supervised by Joe Quatela.
Photocomposed pages prepared from the author's PostScript files.

9 8 7 6 5 4 3 2 1

ISBN 978-1-4612-7525-1　　　ISBN 978-1-4612-2392-4 (eBook)
DOI 10.1007/ 978-1-4612-2392-4

Disclaimer

The programs included in VARIOWIN 2.2 are distributed without any express or implied warranty. Any use of the programs in situations that could result in personal injury or property loss is done at the user's own risk. THE AUTHOR DISCLAIMS ALL LIABILITY FOR DIRECT OR CONSEQUENTIAL DAMAGES RESULTING FROM USE OF THE VARIOWIN PACKAGE.

Trademarks

Several registered trademarks appear within this user's guide:
- Windows™ is a registered trademark of Microsoft Corporation;
- SURFER refers to proprietary computer software of Golden Software Inc.

Contact Addresses

VARIOWIN 2.2 users can e-mail their questions, suggestions, and bugs reports to **variowin@springer-ny.com**. Although the author cannot guarantee he will respond to these messages, he will use them to update the frequently asked questions (FAQ) page that can be accessed on the world wide web at **http://www.springer-ny.com/supplements/variowin.html**.

Contents

1
Introduction

1.1 Overview

VARIOWIN 2.2 is a collection of four Windows™ programs – Prevar2D, Vario2D with PCF, Model, and Grid Display – that are used for spatial data analysis and variogram modeling of irregularly spaced data in two dimensions.

Prevar2D builds a pair comparison file (PCF), that is, a binary file containing pairs of data sorted in terms of increasing distance. Pair comparison files can be built from subsets in order to reduce memory requirements.

Vario2D with PCF is used for spatial data analysis of 2D data. It uses an ASCII data file and a binary pair comparison file produced by Prevar2D. Features implemented in Vario2D with PCF include:

- the possibility to characterize the spatial continuity of one variable or the joined spatial continuity of two variables,
- variogram surfaces for identifying directions of anisotropies,
- directional variograms calculated along any direction,
- several measures of spatial continuity. Not only the variogram but also the standardized variogram, the covariance, the correlogram, and the madogram are used to measure spatial continuity.
- h-scatterplots to assess the meaning of these measures,
- the identification and localization of pairs of data adversely affecting the measure of spatial continuity. Once identified, these pairs can be masked from the calculation interactively.
- variogram clouds for identifying pairs of data values having the most influence on the measure of spatial continuity. Those pairs can also be located on the sample map.
- the ability to save directional variograms in an ASCII file that can be used for subsequent modeling by the Model program.

The Model program is used to produce a 2D nested model of spatial continuity in an interactive way. Several directional variograms read from a variogram file produced by Vario2D with PCF are adjusted simultaneously by a

2D nested model. The adjustment is done with scroll bars. Each time a parameter of the 2D model is changed, cross sections through the 2D model are recalculated and redrawn on the experimental variograms used for the fitting procedure. The 2D nested model can be saved in a model file that is used to store the various parameters required for geostatistical estimation – kriging – or simulation.

Grid Display is used for producing pixel maps of experimental and modeled variogram surfaces that have previously been saved into grid files.

1.2 History

VARIOWIN was initially developed as part of the author's Ph.D. [Pannatier, 1995]. The project started in September 1992 and version 1.0 was presented at an international conference in Bari, Italy, at the end of September 1993 [Pannatier, 1994].

Version 2.0 was selected for the final round of the *European Academic Software Award 1994* which took place in Heidelberg, Germany, at the end of November 1994. Improvements from version 1.0 included those features:
- thorough methodological on-line help,
- sample map and variogram surfaces with correct aspect ratio,
- the ability to select several pairs and locate them on an h-scatterplot and on a variogram cloud,
- the inclusion of zero distance pairs in the calculation of the variogram surface,
- increased ease in building a 2D nested model,
- the capacity to check the same 2D nested model can against several experimental measures of spatial continuity,
- the ability to save a 2D nested model as a grid file, allowing its representation as a variogram surface.

Version 2.1 was made available as a freeware on the world wide web [http://www-sst.unil.ch/geostatistics.html] in March 1995. It is the version that was presented in the Ph.D. thesis. Improvements from version 2.0 included the following:
- the replacement of the standardized inverted (cross) covariance by the standardized (cross) variogram (This new measure of spatial continuity has properties similar to the general relative variogram),
- the ability to handle data sets with UTM coordinates,
- the possibility to modify the parameters of a 2D nested model of spatial continuity not only via scroll bars but also by entering values with the keyboard,
- easier interaction with (cross) variogram clouds,
- the ability to read data files with empty lines at the end of the file,

- enhanced appearance of the various graphs produced by Vario2D with PCF and Model.

Version 2.2 has been developed for this book. Improvements from version 2.1 include:

- a new program, Grid Display, for displaying grid files as pixel maps. Experimental and modeled variogram surfaces can now be compared on-screen.
- the capacity to copy to the clipboard all plots produced by Vario2D with PCF, Model, and Grid Display. Those plots can be copied as bitmaps or as a collection of objects using the Windows metafile format.
- an Install and Uninstall utilities that make it very easy to install/uninstall VARIOWIN 2.2 under Windows 3.x, 95 and NT,
- A completely revised user's guide – this book – that replaces the on-line help.

> Note that version 2.2 is distributed with this book and will not be available on the internet.

1.3 Warning : This Is a Methodological User's Guide

This user's guide has been developed for teaching purposes. Thus, it may resemble a textbook more than a traditional user's guide. The fundamentals of spatial data analysis and of geostatistics are recalled and some suggestions on what to do and what to avoid when using the software for a geostatistical study are given.

1.4 Notation Used Within This User's Guide

Items followed by a number enclosed in brackets, such as "variogram surfaces [4.4]," are defined in the subsection corresponding to the number.

Words written in bold, such as **left click**, indicate an action that can be performed within one of the programs included in VARIOWIN 2.2.

All vectorial quantities, such as a location **x** or a separation vector **h**, are written in bold. |**h**| refers to the magnitude of vector **h**. According to standard math convention, random variables and random functions are denoted with capital letters. Thus $Z(\mathbf{x})$ refers to either the random variable or the random function considered at location **x**.

1.5 Minimum Requirements

The programs contained in VARIOWIN 2.2 will run on any PC with a 386 processor or above running Windows 3.x, 95 or NT. A coprocessor is also required because of the intensive computation required for variogram calculation and variogram modeling. The package requires approximately 2 MB on a hard disk.

1.6 Content of the Release

The following files are included with this release and will be installed into the VARIOWIN directory:

PREVARD2D.EXE	the Prevar2D program,
VARIO2DP.EXE	the Vario2DP with PCF program,
MODEL.EXE	the Model program,
GDISPLAY.EXE	the Grid Display program,
VARIOWIN.HLP	VARIOWIN 2.2. help file,
VARIOWIN.INI	initialization file,
SAMPLES.DAT	contains the medium-sized data set in 2D (470 data) used in Isaaks & Srivastava, 1989,
EXAMPLE.DAT	contains the small-sized data set in 2D (60 data) distributed with the Geo-EAS software [Englund & Sparks, 1991],
GOSSAN2D.DAT	contains the medium-sized data set (547 data) distributed by E. J. Sides, ITC, Kanaalweg 3, 2628 EB Delft, The Netherlands, for a comparative study on variogram modeling. This file contains composites of gold samples from a mine in southern Spain. Hole 236 and 558 have the same easting and northing coordinates.
GOSSAN.DAT	the GOSSAN2D data set with holes 236 and 558 averaged into a new hole. This file contains 546 data which all have a different location in the 2D space determined by the easting and northing coordinates.
UNINSTALL.EXE	the Uninstall utility.

1.7 Installation Procedure

The Install utility – INSTALL.EXE – expands the compressed files on the VARIOWIN 2.2 disk and copies them to the target directory you specify. It also creates the UNINSTALL.LOG file which will be used by the Uninstall program if you ever want to remove the software from your computer.

1.7.1 Note to Windows 95 Users

VARIOWIN 2.2 does not function properly on a system where Microsoft Plus has been installed on top of Windows 95. This problem is not related to Windows 95 but to the system agent installed by Microsoft Plus. This system agent must be stopped prior using VARIOWIN under Windows 95 and Microsoft Plus (go to **Programmes** | **Accessories** | **System tools** | **System agent**, choose the **Advanced** | **stop using system agent** menu item and restart your computer).

1.7.2 Note to Windows NT Users

The Install program should be run by a user who has the right to modify the WIN.INI file that is located in the Windows directory. In order for VARIOWIN users to be able to save their preferences, they must have a write access to the VARIOWIN.INI file located in the VARIOWIN directory (use the **Security** | **Permissions...** menu item in the File Manager program).

1.8 Uninstallation Procedure

VARIOWIN 2.2 can be removed properly from your computer with the Uninstall utility that was copied to your VARIOWIN directory. This program uses the UNINSTALL.LOG file which contains a log of all steps performed during the installation procedure. It should be run by a user who has sufficient privileges to remove the VARIOWIN directory and modify the WIN.INI file located in the Windows directory.

Acknowledgments

The author acknowledges R. Froidevaux, from FSS International in Geneva, and M. Maignan, M. Jaboyedoff, and L. Calmbach, from the Institute of Mineralogy of the University of Lausanne, for their useful suggestions while the software was being developed.

References

Englund, E. & Sparks, A., "Geo-EAS 1.2.1 User's Guide," US-EPA Report #600/8-91/008, EPA-EMSL, Las Vegas, NV, 1991.

Isaaks, E. H. & Srivastava, R. M., "An Introduction to Applied Geostatistics," Oxford University Press, New York, NY, 1989.

Pannatier, Y., "VARIOWIN: Logiciel pour l'analyse spatiale de données en 2D – Etude géologique et géostatistique du gîte de phosphates de Taïba (Sénégal)," Ph.D. thesis, University of Lausanne, Lausanne, Switzerland, 1995.

Pannatier, Y., "MS-WINDOWS Programs for Exploratory Variography and Variogram Modeling in 2D," in *Statistics of Spatial Processes: Theory and Applications*, Capasso,V., Girone, G. & Posa, D. (eds.), Istituto per Ricerche de Mathematica Applicata (IRMA), Bari, Italy, pp. 165–170, 1994.

2
Quick Start

In this chapter the four programs included in VARIOWIN 2.2 are used for the spatial analysis of a 2D data set. The data files gossan2d.dat and gossan.dat are used for this demonstration. They are included with this release of VARIOWIN, and so you may repeat the exercise as a tutorial, or as a test for the software.

The goal here is to characterize and model the spatial continuity of the variable gold. The following screen captures illustrate how the software was used for this study.

2.1 Prevar2D – Construction of a Pair Comparison File

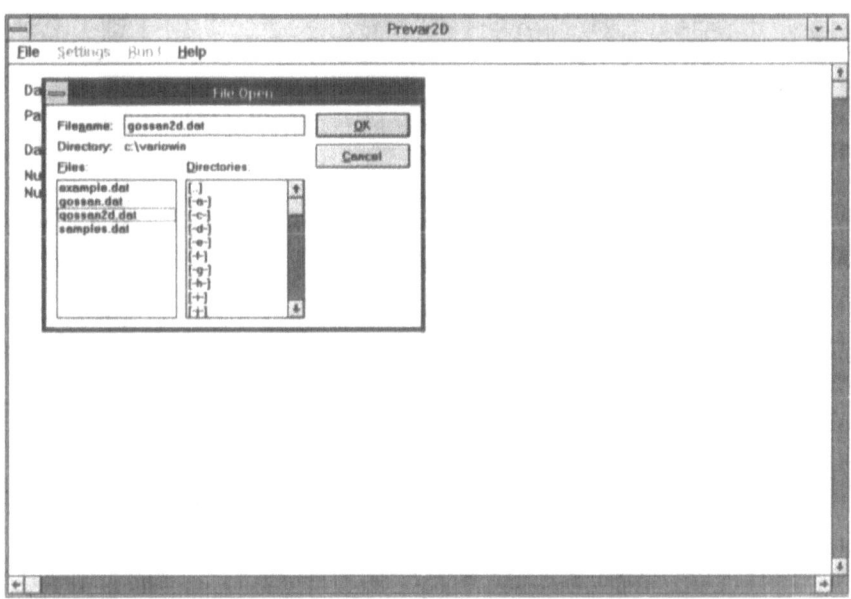

Step 1 Load the data file gossan2d.dat (use the **File | Open Data File...** menu item).

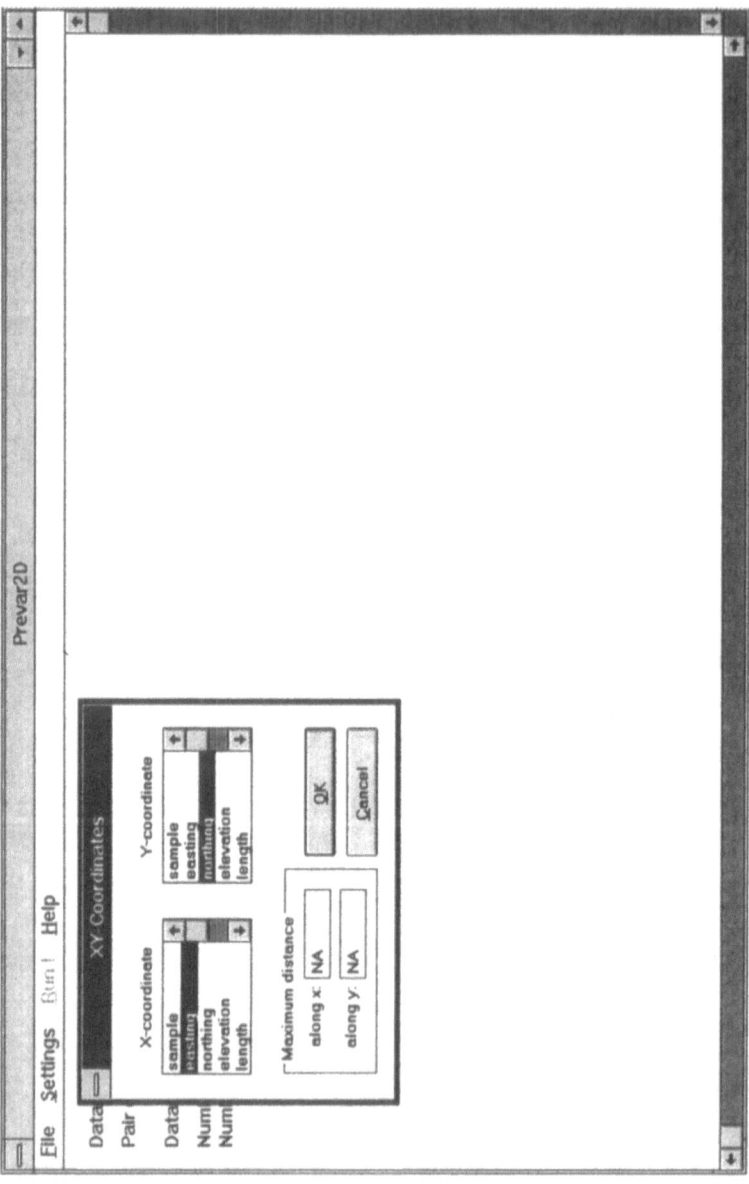

Step 2 Choose the **Settings | XY-Coordinates...** menu item and the appropriate *x*- and *y*-coordinates, that is, easting and northing. Try to build a pair comparison file with the **Run !** menu item which is enabled once the *x*- and *y*-coordinates have been set.

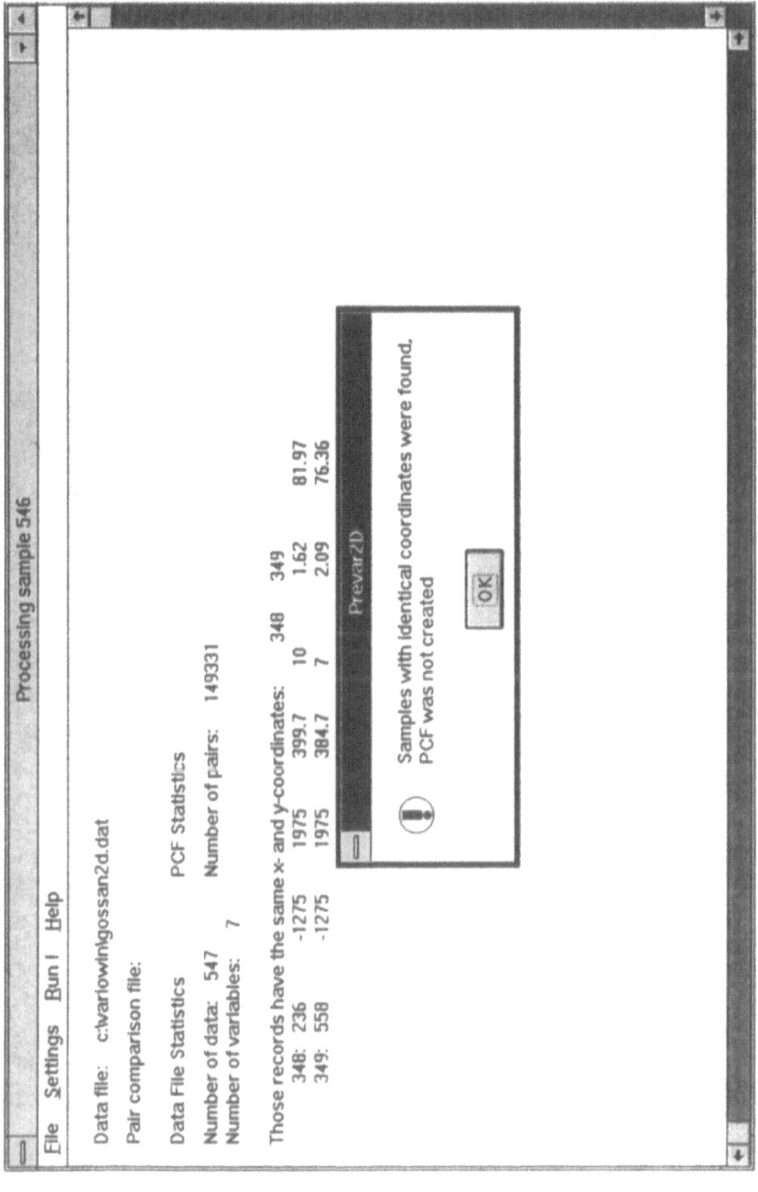

Step 3 This is what results when samples with identical coordinates have been found. No pair comparison file is created, and the first 20 samples having identical x- and y-coordinates are displayed.

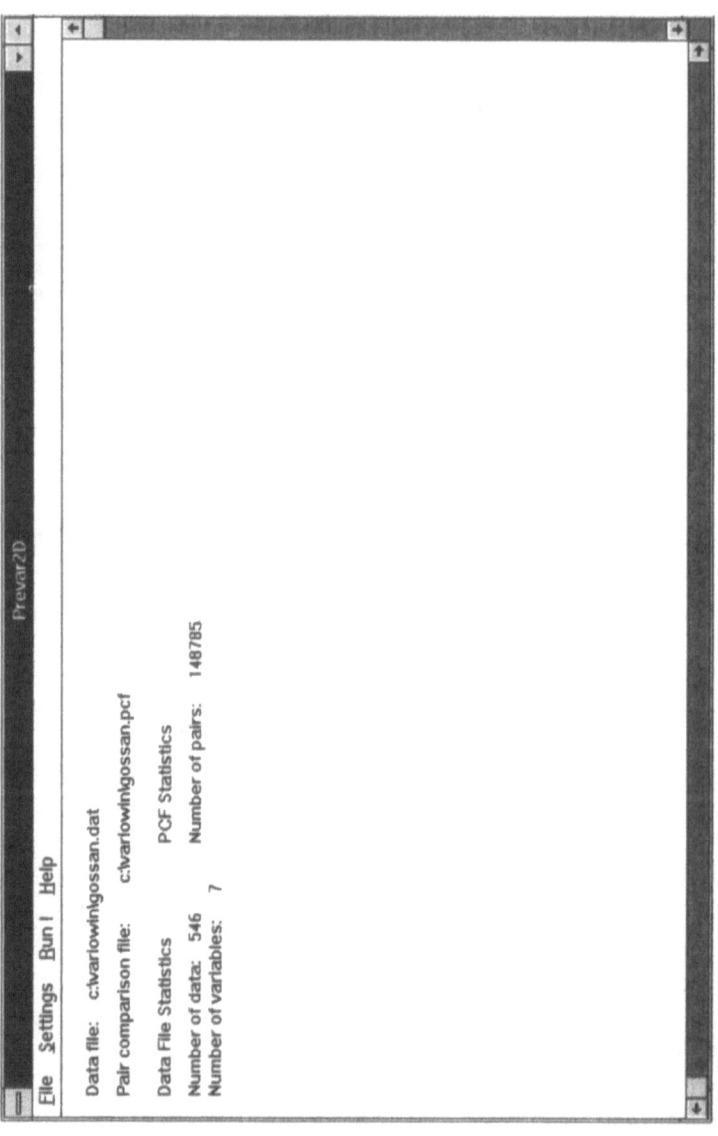

Step 4 In data file gossan.dat, holes 236 and 558 – records 348 and 349 – have been manually averaged into a new hole. A pair comparison file can be created using this data file which does not contain any duplicate point (note that the x- and y-coordinates should be set equal to easting and northing after gossan.dat has been read into memory).

2.2 Vario2D with PCF – Exploratory Variography

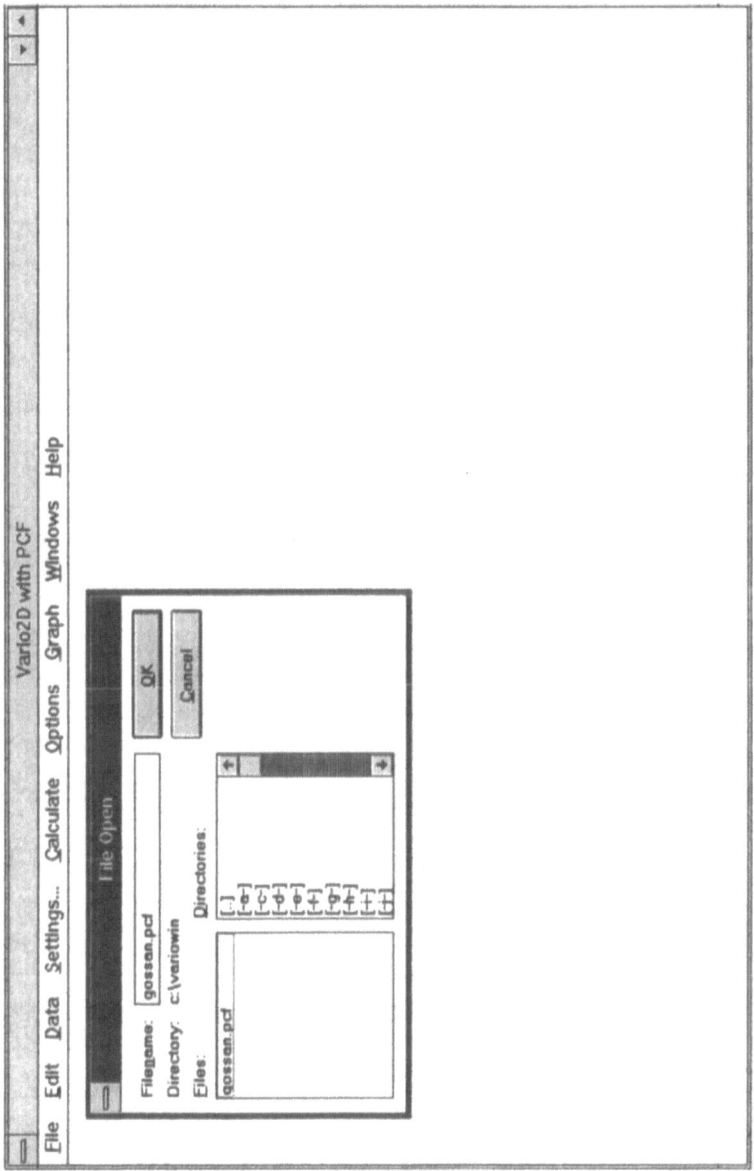

Step 1 Load gossan.pcf, the pair comparison file created by Prevar2D. Note that Vario2D with PCF expects to find the original data file, gossant.dat, and the pair comparison file, gossan.pcf, in the same directory.

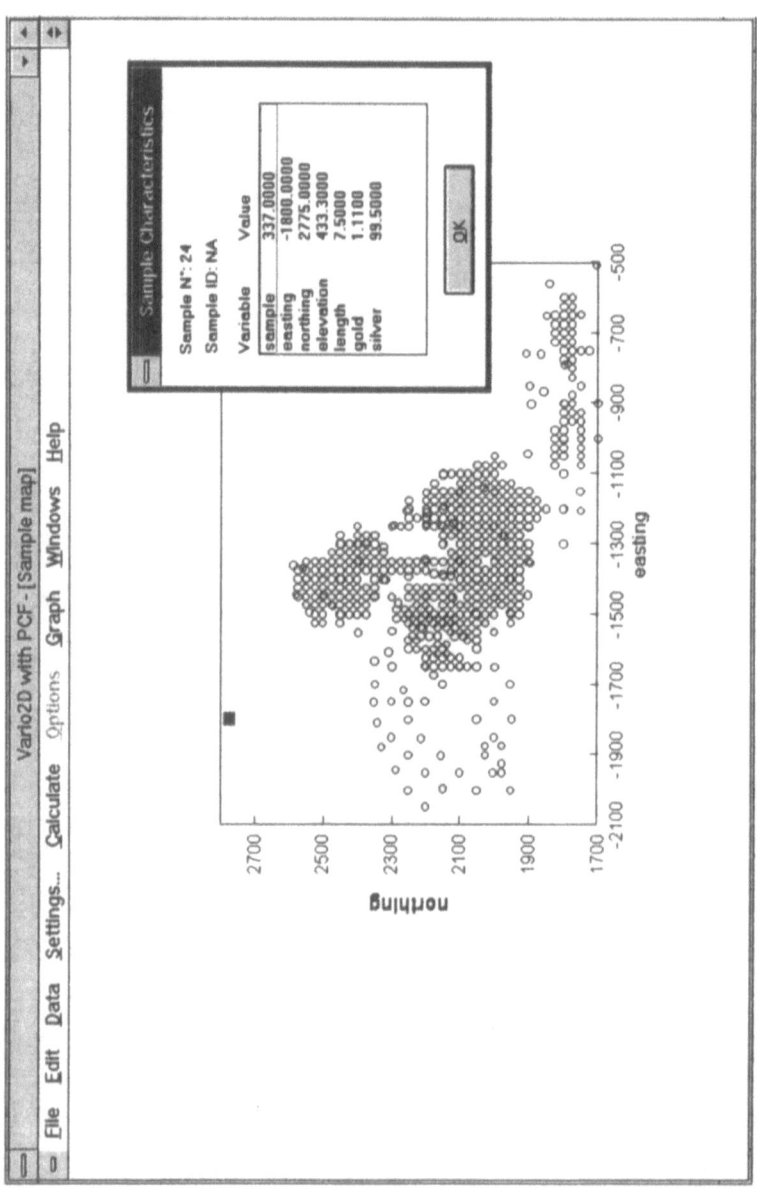

Step 2 You can now display a map of the data (use the **Data | Map !** menu item (shortcut **F10**)) and display the characteristics of a sample by clicking on it. Samples numbers can be displayed using the **Graph | Records** menu item (shortcut **F6**) when the sample map is the active window.

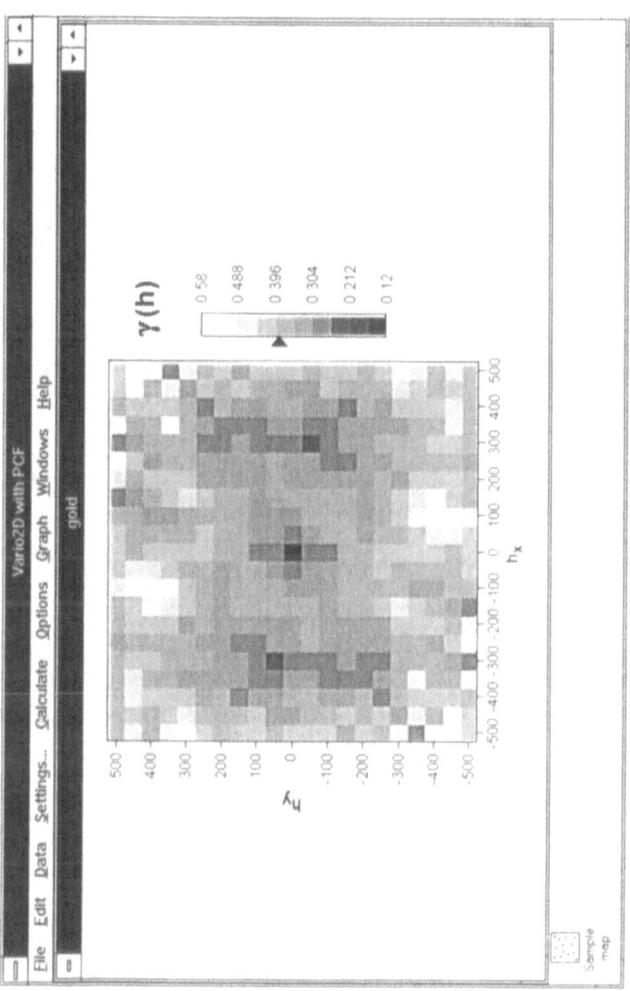

Step 3 The calculation of a variogram surface (use the **Calculate | Variogram Surface...** menu item, set the lag spacing and the number of lags in direction X and Y to 50 and 10) allows identification of the main directions of anisotropies of the spatial behavior of a variable. The variogram surface shows that the variable gold has a maximum continuity in direction 40 and a minimum continuity in direction 130 (trigonometric angle). Different measures of spatial continuity can be displayed (use the **Options | Estimator** menu item (shortcuts **F1** to **F5**)). The graph's appearance can be modified using the menu items found under the **Graph** menu. The variogram surface for gold should be saved in the file expgold.grd which will be used later in the demonstration (use the **File | Save as...** menu item when the variogram surface is the active window).

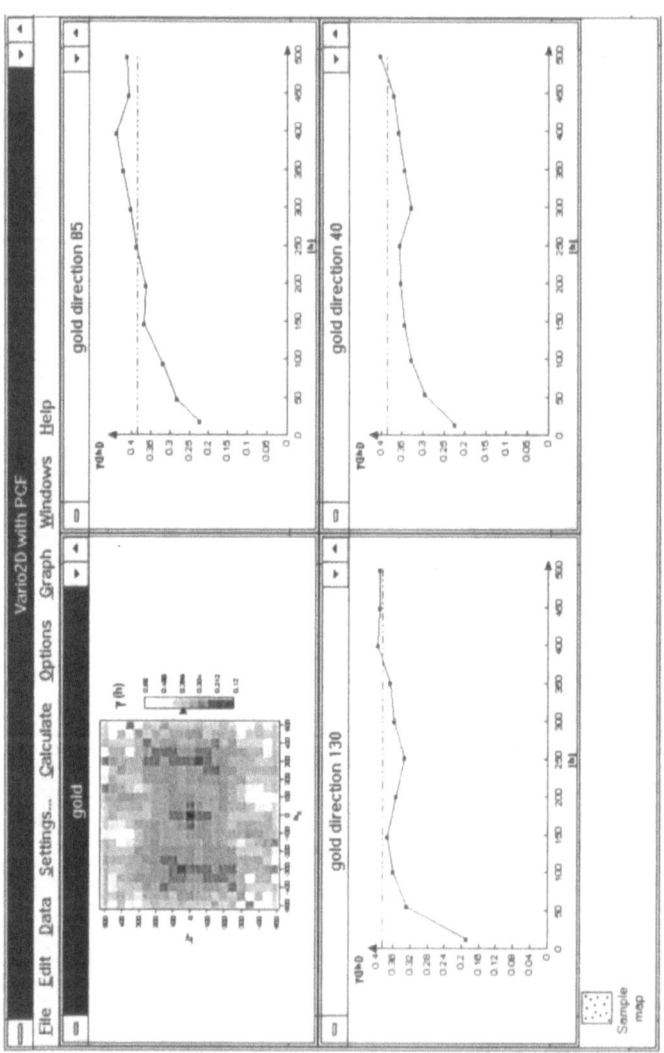

Step 4 Experimental variograms can be calculated along the directions identified on the variogram surface (use the **Calculate** | **Directional Variogram...** menu item, choose the variable gold and the direction, set the lag spacing to 50, the number of lags to 10 and the angular tolerance to 30). Variograms can be tiled on-screen using the **Windows** | **Tile** menu item. Different measures of spatial continuity can be displayed (use the **Options** | **Estimator** menu item (shortcuts **F1** to **F5**)). The directional variogram's appearance can be modified using the menu items found under the **Graph** menu.

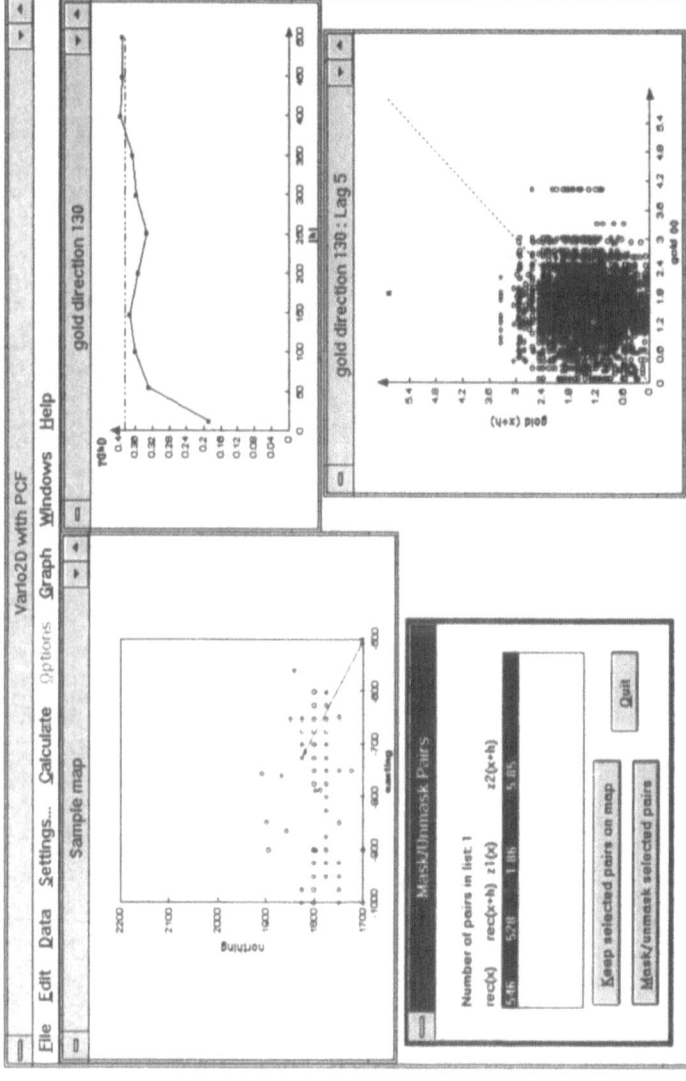

Step 5 The h-scatterplot corresponding to lag 5 of the directional variogram calculated in direction 130 has been displayed (just click on the point in the variogram window). The isolated pair on the h-scatterplot has been clicked and the program displays it on the sample map. It is also possible to keep it on the map by using the **Keep selected pairs on map** button. The sample map's zoom was obtained using the **Graph | Axis...** menu item.

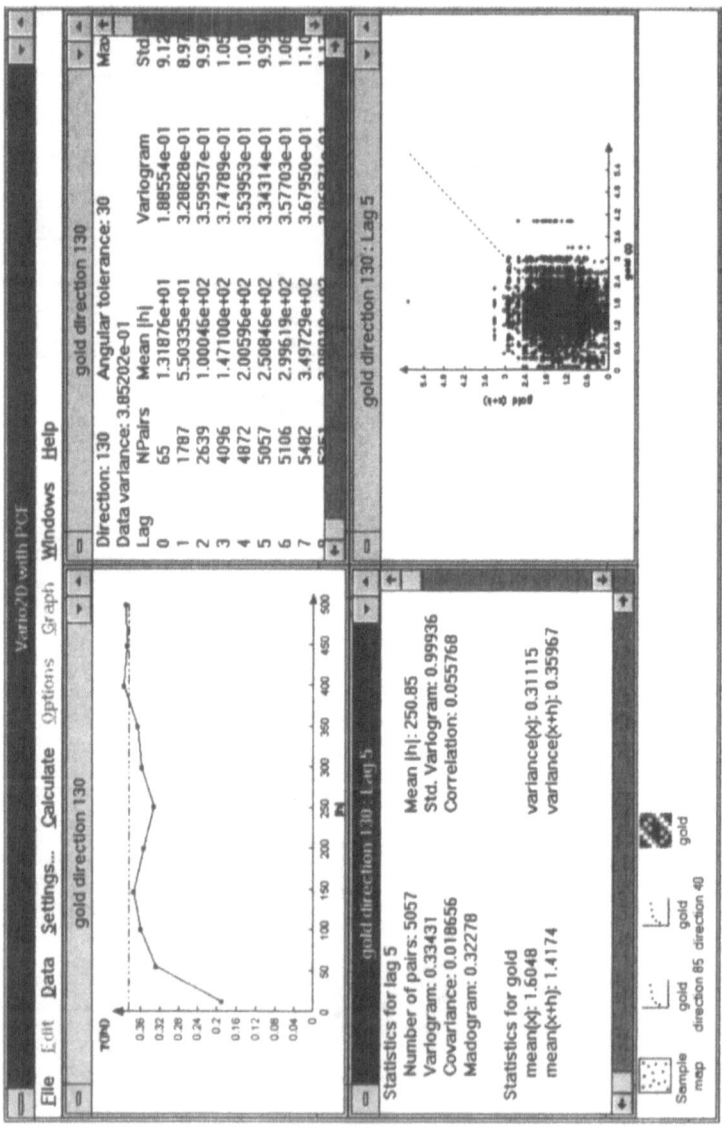

Step 6 Numerical results for the directional variograms in direction 130 and for the h-scatterplot corresponding to the fifth lag have been displayed along with their graphical counterparts (use the **Options | Numerical Results !** menu item (shortcut **right click** anywhere in the variogram or the h-scatterplot window)).

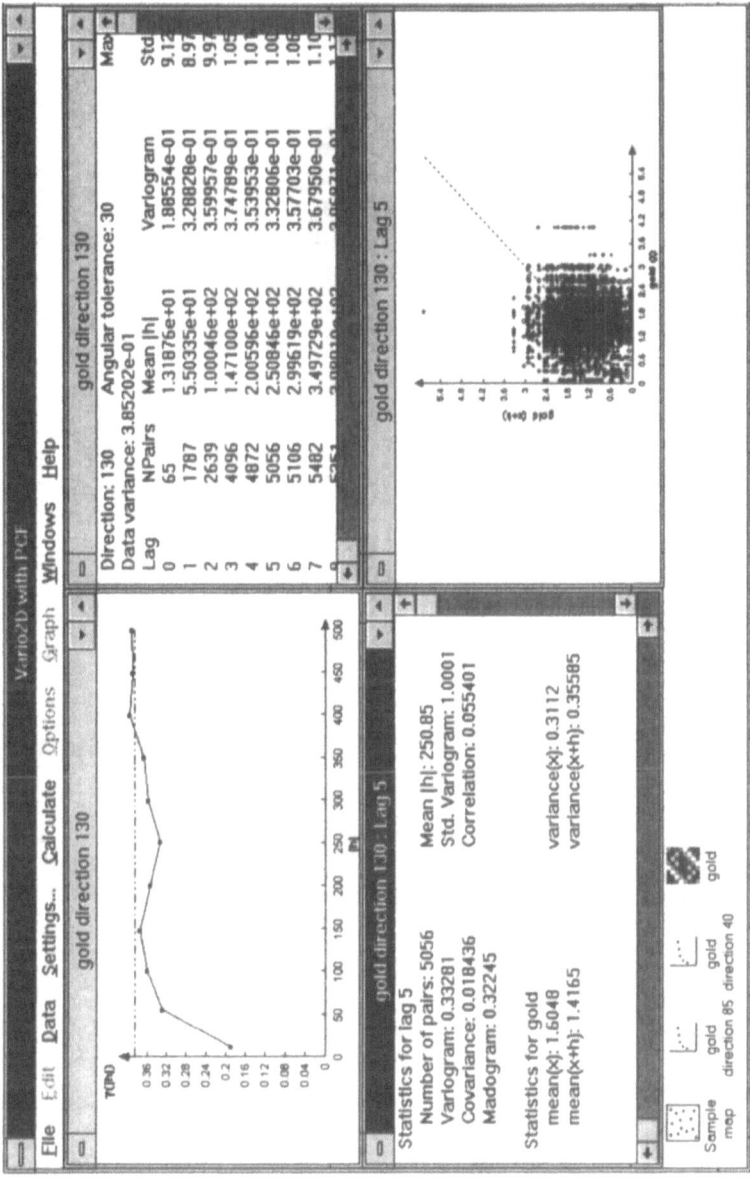

Step 7 The isolated pair on the h-scatterplot has been masked. Note that all statistics relative to the fifth lag have been updated as well as those of the corresponding directional variogram.

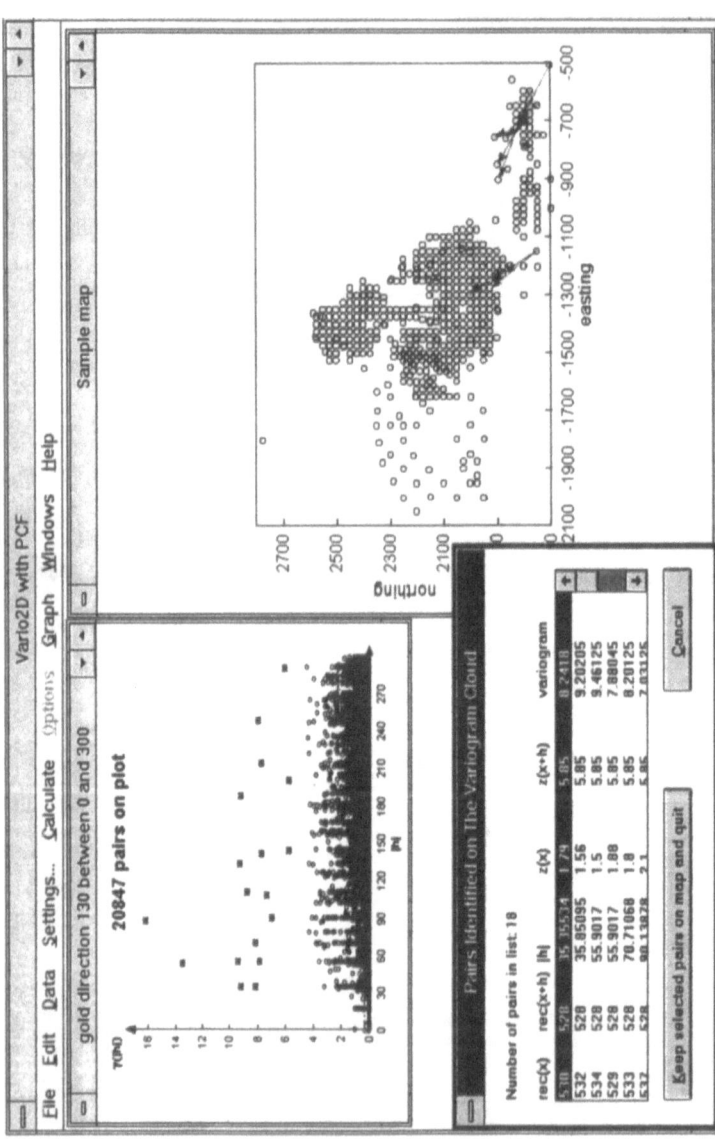

Step 8 The variogram cloud in direction 130 has been calculated (use the **Calculate | Variogram Cloud...** menu item, choose the variable and the direction, set the maximum distance to 300 and the angular tolerance to 30). The pairs with the biggest squared differences have been identified (use **ctrl + left click** and **left click** to end the selection) and displayed on the sample map.

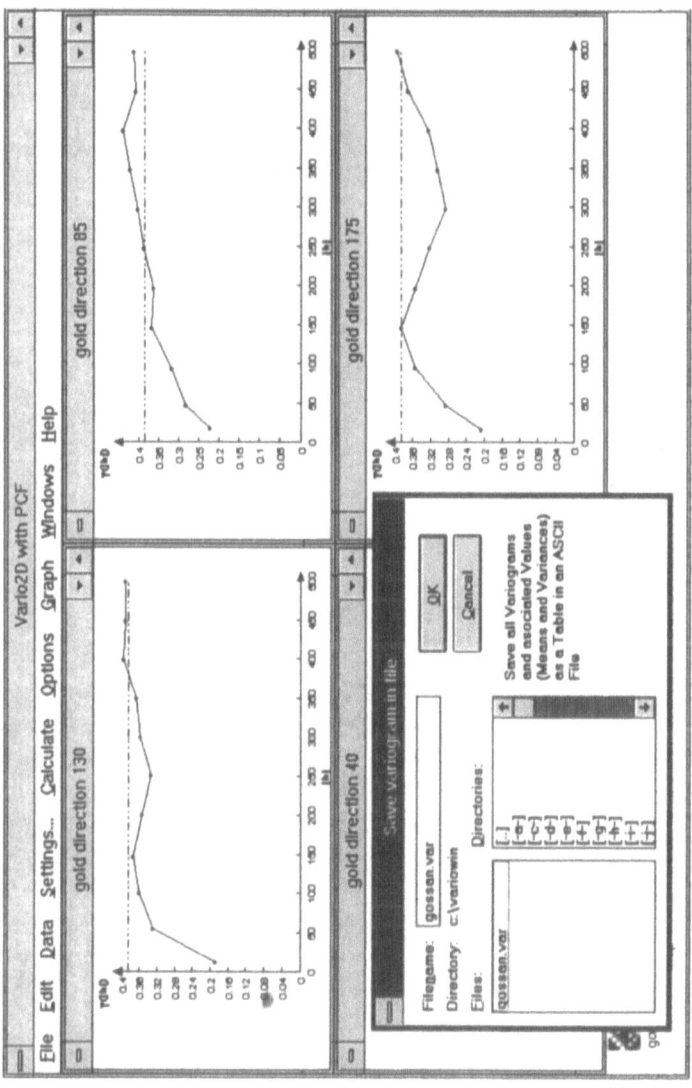

Step 9 Once the directional variograms have been analyzed and are deemed representative of the spatial continuity of the studied variable, they can be saved in a variogram file (use the **File | Save as...** menu item) for subsequent modeling. Note that several directional variograms can be saved in the same file (just pick up an existing file and click on the **Append** button). The four directional variograms shown on the screen capture should be saved to the file gossan.var which will be used in the remaining part of this chapter.

2.3 Model – Interactive Variogram Modeling

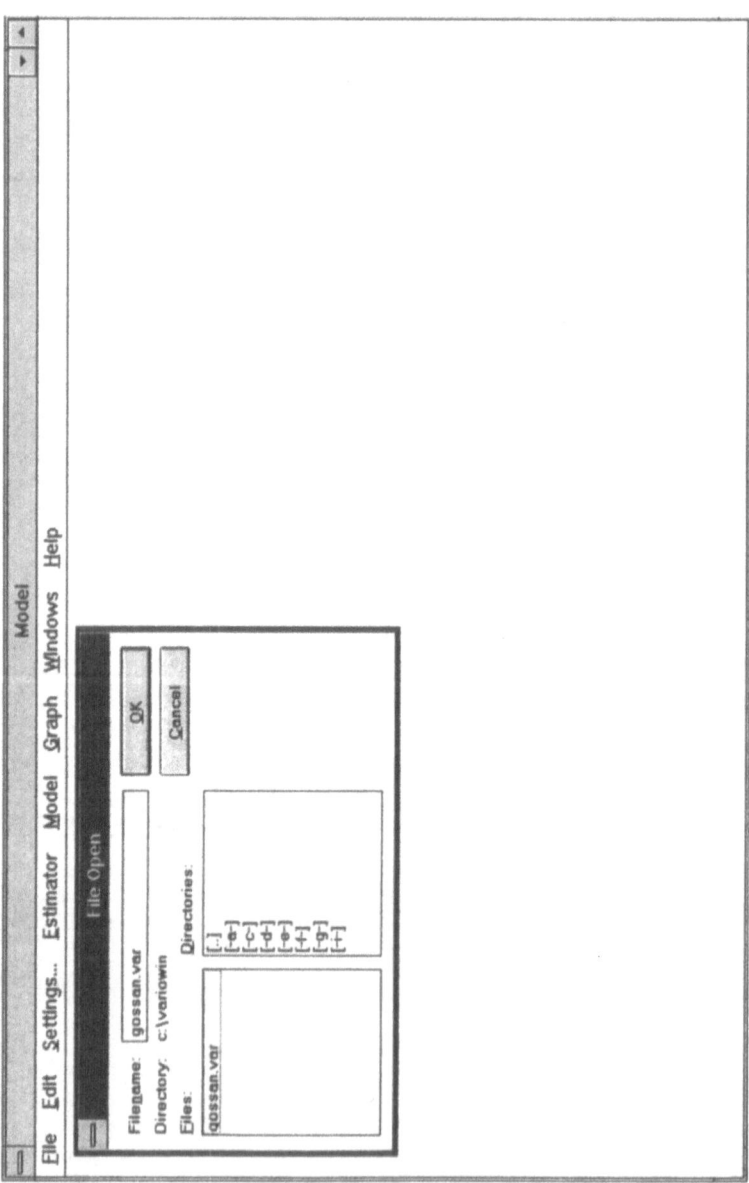

Step 1 Load the variogram file – gossan.var – created by Vario2D with PCF. This file contains all experimental variograms that will be needed for the modeling session.

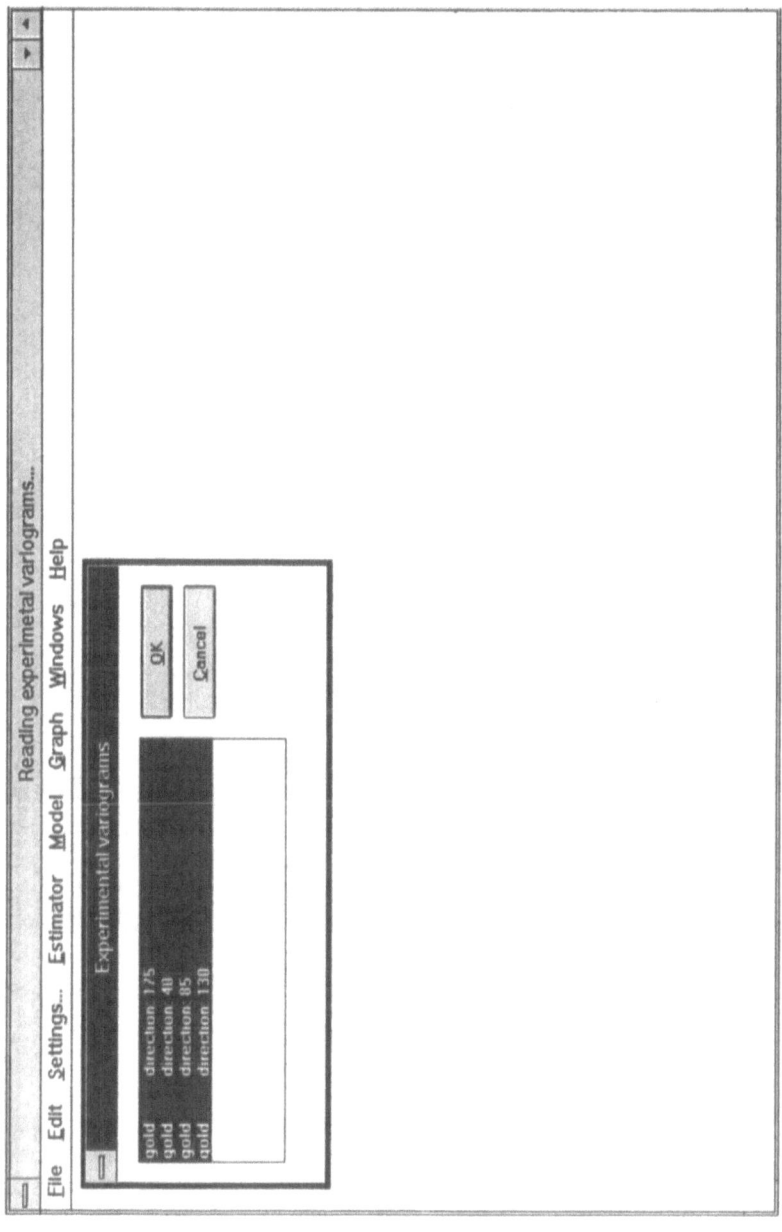

Step 2 Choose which experimental variograms you want to use for fitting the 2D nested model.

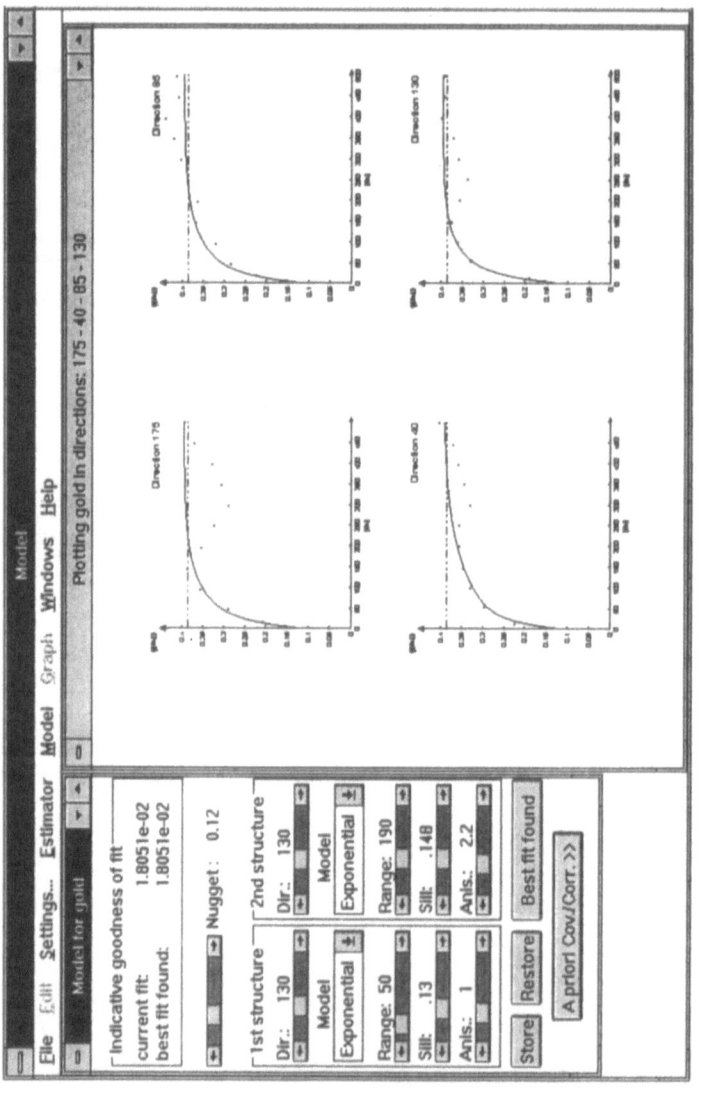

Step 3 Adjust the 2D nested model parameters with scroll bars or by entering your own values with the keyboard. Watching how the modeled variograms change as the parameters of the 2D model are modified, you should try to obtain a model that fits all experimental variograms as well as possible. This 2D model should be saved as a variogram surface in the file modgold.grd (use the **File | Save as...** menu item when the plot window is the active window and set the lower limit, the upper limit and the spacing in the *x*- and *y*-direction to respectively -500, 500 and 25).

2.4 Grid Display – Displaying Grid Files as Pixel Maps

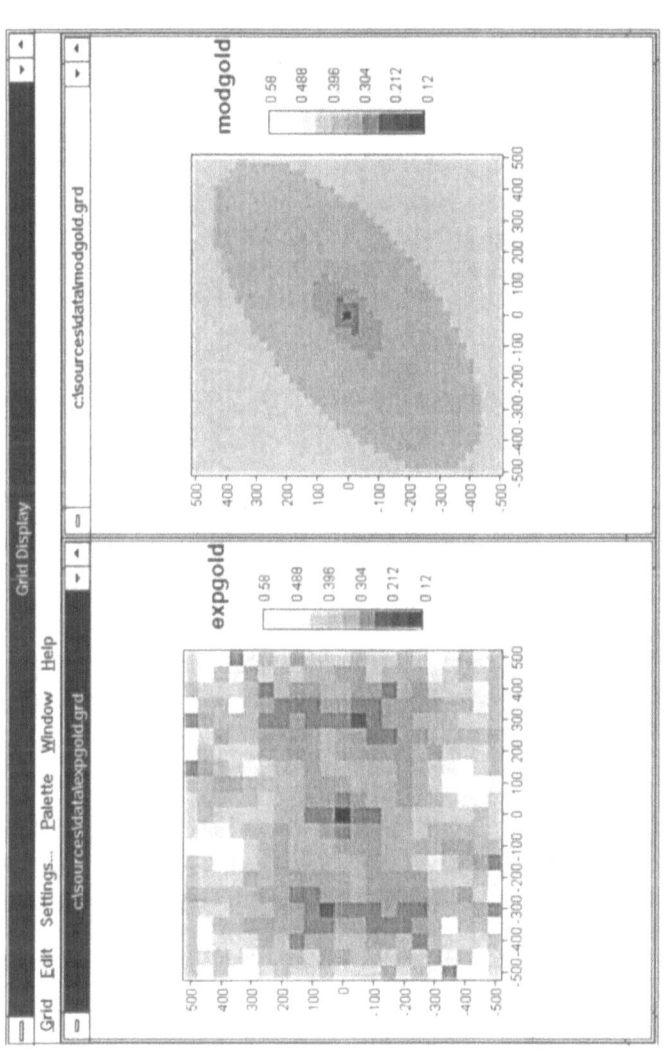

The experimental variogram surface that was stored in expgold.grd and the modeled variogram surface just saved in modgold.grd can now be compared using the Grid Display program (use the **Grid | Open...** menu item to display a grid file). Note that the minimum and maximum gold values in both grid files must be the same (see paragraph 6.3). The 2D nested model of spatial continuity reproduces the main anisotropy observed on the experimental variogram surface, but it does not reproduce the hole effects visible in directions 40, 130, and 175.

3

Construction of a Pair Comparison File (PCF) with Prevar2D

3.1 Overview

Prevar2D constructs a PCF from data files that can hold any number of samples provided enough memory is available on the computer. This pair comparison file contains the number of pairs calculated, the sorted list of pairs, and the name of the data file from which the PCF was constructed.

A PCF can be constructed with the **Run !** menu item, which is enabled after the user has validated the "ettings" dialog box displayed with the **Settings | XY-Coordinates...** menu item. This menu item is enabled when a data file [6.1] has been read into memory (use the **File | Open Data File...** menu item).

Prevar2D builds a pair comparison file in two steps:
1. All pairs belonging to the active subset [3.4] are first written to a binary file.
2. This binary file is then loaded into memory and pairs are sorted by increasing distance using the quicksort algorithm [Press et al., 1992]. At this stage an error message can be displayed if the memory available on the system is not sufficient to load the binary file. A possible solution is to increase the swapping space used by Windows.

3.2 What Is a PCF?

A pair comparison file (PCF) is a binary file that contains pairs of samples sorted by distance. It also includes either a copy of the original data file or a reference to it.

Various public domain packages, such as Geo-EAS [Englund & Sparks, 1991] or the Geostatistical Toolbox [Froidevaux, 1990], make use of a PCF. However, those programs work under DOS, and the number of pairs they can handle is limited to about 18,000 (190 data). On the other hand, Prevar2D can handle any number of pairs, provided there is enough memory on the system. The author has worked with more than 500,000 pairs on a 486/66 PC with 8 MB of memory without any problem.

3.3 Working with a PCF

3.3.1 Advantages

1. Calculating a measure of spatial continuity [4.7] is fast with the help of a PCF since all pair distances have already been computed.
2. Direct access to the pairs is granted; this allows the construction of h-scatterplots [4.3.1] and variogram clouds [4.6].

3.3.2 Problems

1. This approach is not practical when dealing with regularly spaced data since all distances can easily be derived from the grid layout.
2. The number of pairs can become very large:

$$\left(\frac{\text{Nsamples} \cdot (\text{Nsamples} - 1)}{2} \right),$$

which makes it difficult to sort the vector of pairs into memory. However, the Windows environnement allows Prevar2D to handle large arrays of pairs.
3. For large data sets, those with several thousand of data, the time spent searching and reading the pairs in the PCF exceeds the time needed to calculate the measure of spatial continuity from original data.

3.3.3 Overcoming DOS Memory Limitations

Windows 3.1 can address up to 256 MB of linear memory using the memory manager provided with MS-DOS 6.0, whereas DOS can only address 640 K of linear memory.

With a 16 bits memory architecture, and in order to correctly address a vector of pairs encompassing several 64 K segments, care must be taken in designing the structure containing the pair information: its size must be a factor of 64 K [Microsoft Corporation, 1992].

Structure *Pair2D*, that is used to represent a pair in VARIOWIN 2.2, contains the following information:

Information	Type	Size [bytes]
record(x): position in data file of the tail record	int	2
record(x + h): position in data file of the head record	int	2
|h|: pair's distance	float	4
h_x : difference in x-coordinates	float	4
h_y : difference in y-coordinates	float	4
size of structure		16

3.4 Building a PCF from a Subset

In order to reduce the size of the pair comparison file, and hence the memory requirements of Prevard2D and Vario2D with PCF, a pair comparison file can be constructed from subsets of the available data:

1. Subsets can be based on maximum difference in x- and/or y-coordinates using the **Settings I XY-Coordinates...** menu item.
2. Subsets can be based on acceptable ranges of variables using the **Settings I Limits...** menu item.

3.5 What Should Be Done with Duplicate Data Points?

Prevar2D cannot build a PCF when several samples have the same X- and Y-coordinates. Consequently, when several values of a variable are available at a specific location, the data file must be manually edited. Records having identical x- and y-coordinates, but different values of the variable(s) of interest, should be replaced with one record where the variable(s) is/are a statistic of the original values. The median, mean, mode, minimum, or maximum should be considered.

Further Reading

Englund, E. & Sparks, A., "Geo-EAS 1.2.1 User's Guide," US-EPA Report #600/8-91/008, EPA-EMSL, Las Vegas, NV, 1991.

Froidevaux, R., "Geostatistical Toolbox Primer, version 1.30," FSS International, Chemin de Drize 10, 1256 Troinex, Switzerland, 1990.

Microsoft Corporation, "Microsoft Windows 3.1 – Guide to Programming", Microsoft Press, Redmont, WA, 1992.

Press, W. H., Teukolsky, S. A., Vetterling, W. T. & Flannery, B. P., "Numerical Recipes in C – The Art of Scientific Computing – Second Edition," Cambridge University Press, Cambridge, 1992.

4
Vario2D with PCF – A Program for Interactive Exploratory Variography

4.1 Overview

Vario2D with PCF performs exploratory variography on a 2D data set with the help of a pair comparison file [6.2] constructed with Prevar2D. Note that the program expects to find the data file [6.1] and the pair comparison file in the same directory.

Because Vario2D with PCF is a multiple-document interface (MDI) compliant application, several variographical views of the data can be tiled on-screen, allowing the simultaneous examination of spatial continuity from several points of view. Figure 4.1 illustrates the content of those views and their relationships:

1. The *sample map* is used to identify errors in the coordinates and data clustering. Pairs of data are also plotted on this map.
2. An *h-scatterplot* is the bivariate equivalent of the histogram. In the same way that a histogram is an approximation of the underlying univariate *probability density function* that characterizes the studied phenomenon, an h-scatterplot is an approximation of the underlying bivariate probability density function that characterizes the spatial continuity for a separation vector **h**.
3. The *variogram surface* is an effective way to detect anisotropies in the pattern of spatial continuity. Each cell of a variogram surface represents a measure of spatial continuity that summarizes an h-scatterplot. This diagram is used to identify preferential directions in which directional variograms should be calculated.
4. A *directional variogram*, or *experimental variogram*, displays the pattern of spatial continuity in a given direction. It is a cross section through a variogram surface. Each point of the variogram represents a measure of spatial continuity that summarizes an h-scatterplot. In order for the variogram to be a good representation of spatial continuity in a given direction, each one of its points must be a

meaningful summary of its associated h-scatterplot. The construction of experimental variograms representing the pattern of spatial continuity in several directions is a fundamental step in any geostatistical study since those variograms are used to build a 2D model of spatial continuity.

5.   The *variogram cloud* shows the relationship between the magnitude of the pair separation vector and the variogram value of this pair [4.7.1]. A directional variogram can be considered as the moving average of a variogram cloud.

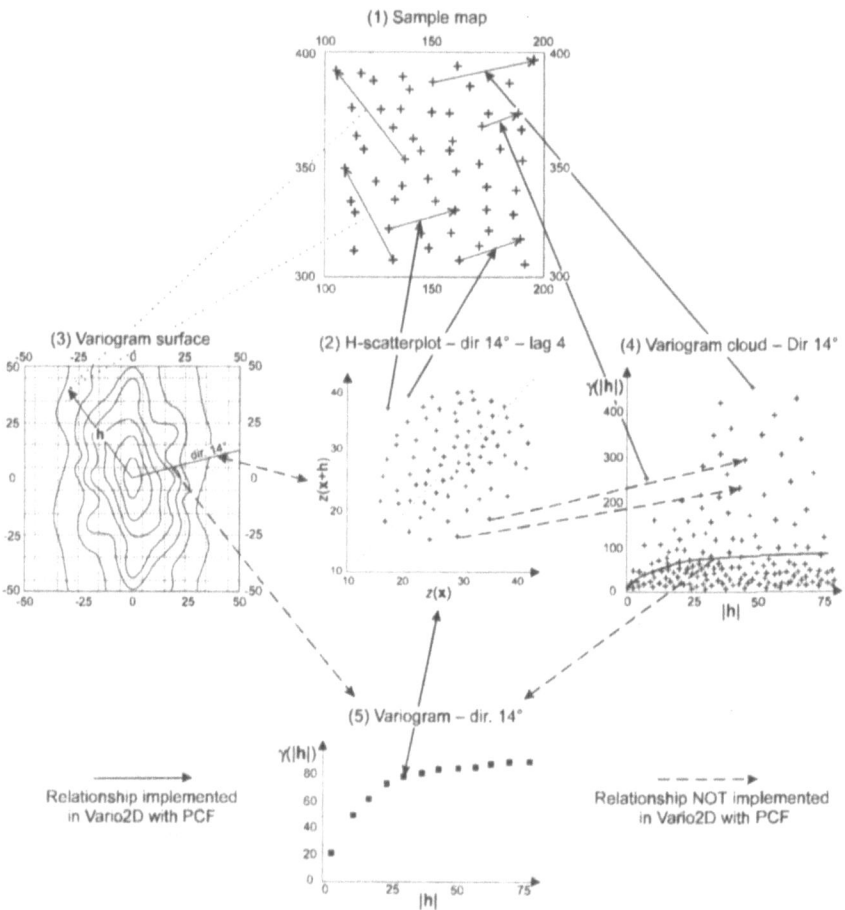

Figure 4.1 Various views of spatial continuity and their relationships.

4.2 Working with Subsets

An active subset can be determined using the **Data | Limits...** menu item. Variogram surfaces, directional variograms, and variogram clouds are calculated using only the data belonging to the active subset. Lower and upper limits can be set for each variable, and a valid range using those limits is associated to each variable. The active subset contains only those samples for which all variables belong to their respective valid range.

The active subset can be visualized on the sample map using the **Data | Map !** menu item. The **F10** function key provides a convenient shortcut for this command. On this sample map, active data will appear with a symbol different from the one used for inactive data. Moreover, interactive identification of samples (**left click** on the sample) is enabled only for data points belonging to the active subset.

4.3 H-Scatterplots and Cross H-Scatterplots

H-scatterplots provide an effective way to grasp the notion of spatial continuity. On an h-scatterplot the value of one variable at position x is plotted against the value of the same variable at position $x + h$, where h is a separation vector. This is referred to as *direct variography*. On a cross h-scatterplot the value of one variable at position x is plotted against the value of another variable at position $x + h$. This is referred to as *cross variography*. Figure 4.2 illustrates the difference between direct and cross variography.

In Vario2D with PCF the choice between direct and cross variography is made in the dialog box displayed with the **Settings...** menu.

When dealing with regularly spaced data, h-scatterplots are constructed according to the procedure illustrated in Figure 4.3.

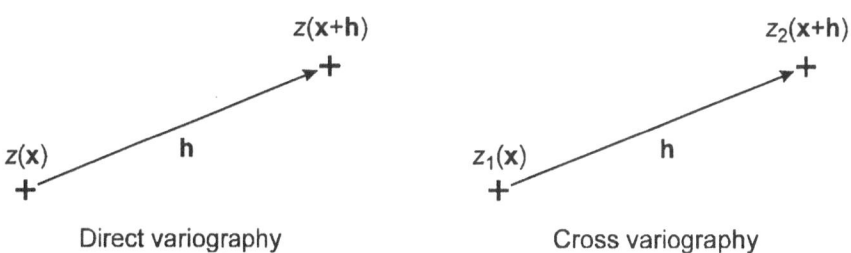

Figure 4.2 Direct and cross variography.

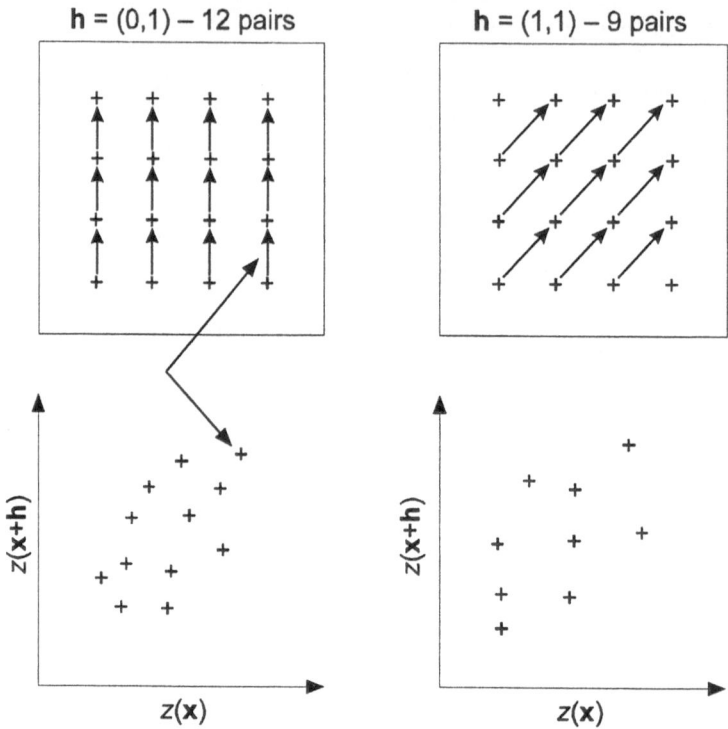

Figure 4.3 Constructing h-scatterplots with regularly spaced data.

If the data are irregularly spaced, all pairs having a separation distance close to the separation vector **h** will be retained to construct an h-scatterplot. In practice, a tolerance is set on the separation vector **h**. Figures 4.4 and 4.5 show how this tolerance can be set:

- on the head of the separation vector **h** (Figure 4.4). This type of tolerance is used to construct variogram surfaces.
- on the distance and on the direction of the separation vector **h** (Figure 4.5). This type of tolerance is used to construct directional variograms.

4.3.1 Working with H-Scatterplots

In Vario2D with PCF h-scatterplots for a given separation vector are displayed when the user clicks on the point of the directional variogram. All statistics pertaining to that separation vector can be presented in a table format using the **Options | Numerical Results !** menu item or by doing a **right click** anywhere in the h-scatterplot window.

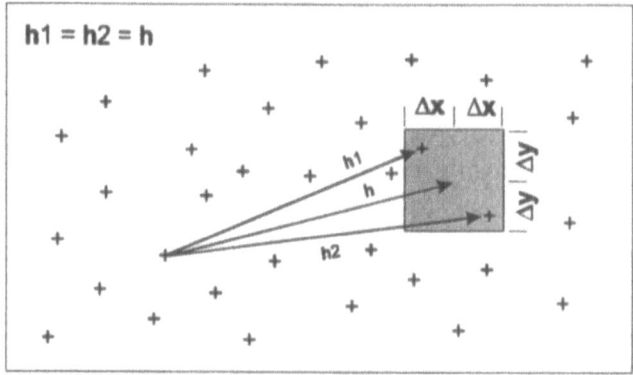

Figure 4.4 Tolerance is set on the head of separation vector **h**.

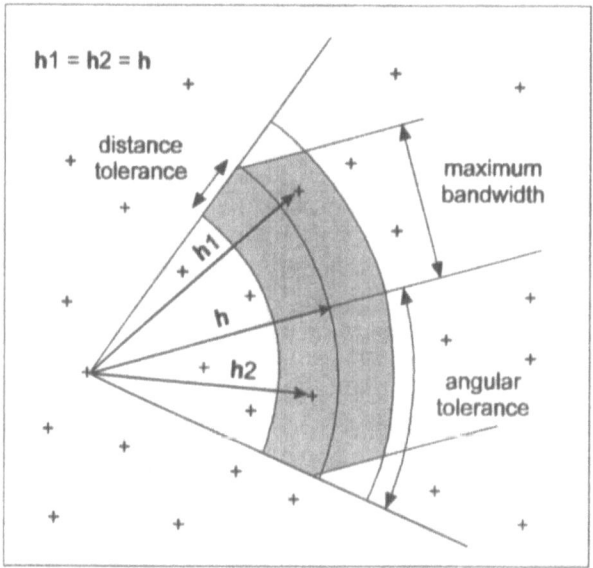

Figure 4.5 Tolerance is set on the direction and the magnitude of separation vector **h**.

H-scatterplots provide an assessment of the true meaning of a directional variogram. In order for a directional variogram to be meaningful, the measures of spatial continuity [4.7] associated with its h-scatterplots must be meaningful:

1. When the h-scatterplot displays a "butterfly-wing" shape (Figure 4.6A), it cannot be adequately summarized with a single number – a measure of spatial continuity – because of its shape. Working with a transformation of the data, such as the logarithmic transform, the normal score transform, or the rank transform, can help. H-scatterplots

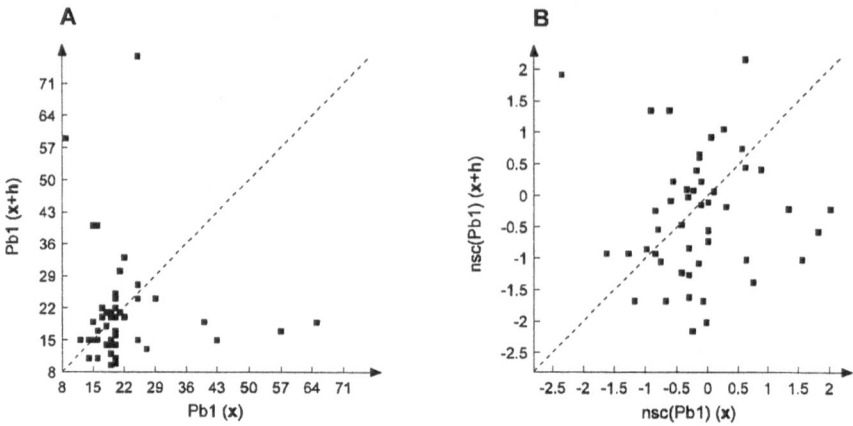

Figure 4.6 H-scatterplot of the original variable (**A**) and h-scatterplot of the normal score transform of the original variable (**B**).

describing the spatial continuity of those transformed variables are usually more symmetric and can be summarized with a single number. Figure 4.6A was constructed using the original data and displays a characteristic butterfly-wing shape. No meaningful measure of spatial continuity can be calculated on such an h-scatterplot. On the other hand, the h-scatterplot in Figure 4.6B was constructed using normal scores of the original data and can be adequately summarized with a single statistic (e.g., the coefficient of correlation, the covariance, or the moment of inertia around the diagonal line).

2. When the data set is a mixture of several populations, its h-scatterplots, calculated for several values of the separation vector **h**, will often show various groups of pairs. This type of h-scatterplot cannot be summmarized adequately with a single number, and the spatial study must be performed separately on each one of the populations involved in the mixture.

H-scatterplots can be used to identify pairs that have a strong influence on a measure of spatial continuity. Pairs selected on an h-scatterplot can be masked (use **Ctrl + left click** to select multiple pairs and **left click** to end the selection). The associated measures of spatial continuity and the parent directional variogram are immediately updated. This interactive masking of pairs and simultaneous recalculation of summary statistics provides an effective way to check the robustness of spatial continuity measures. In order to be robust, a measure of spatial continuity should not change significantly when a small number of pairs located at the periphery of the h-scatterplot are masked. The **Options | Unmask Pairs !** menu item (function key **F9**) can be used to unmask all masked pairs of the active h-scatterplot or to reset all h-scatterplots relative to the active directional variogram.

If the sample map is on-screen (either as an active window or as an icon), identified pairs are also displayed on the map. This interactive identification and posting of anomalous pairs is extremely useful for detecting problems with the data set. This feature can easily detect peculiar points as well as peculiar regions.

H-scatterplots are also useful for checking various assumptions that are frequently made in geostatistics:

1. H-scatterplots of a univariate Gaussian variable provide a quick visual check for the multi-Gaussian hypothesis [A.6]. If the random function representing the phenomenon is multi-Gaussian, all h-scatterplots are elliptical clouds around the diagonal line with a higher density of points in the center of the cloud.

2. A succession of h-scatterplots, calculated for increasing values of the separation vector **h**, provides a check for stationarity [A.4]. If such a succession of h-scatterplots shows that the center of the cloud of pairs departs from the diagonal line, then the random function representing the studied phenomenon cannot be considered stationary.

3. H-scatterplots provide a way to separate different populations. Clouds of points far away from the diagonal line correspond to pairs constructed with points originating from the two populations.

4.4 Variogram Surface and Cross Variogram Surface

In Vario2D with PCF a (cross) variogram surface is produced using the **Calculate I Variogram Surface...** menu item and constructed with the parameters shown in Figure 4.7.

Note that the central cell corresponding to the separation vector (0, 0) includes all pairs with a zero distance. When this cell contains only zero-distance pairs, its measure of spatial continuity reduces to the appropriate statistical estimate:

- 0 for the (cross) variogram [4.7.1], the (cross) madogram [4.7.5], and the standardized (cross) variogram [4.7.2];
- the variance for the covariance [4.7.3];
- the covariance for the cross covariance [4.7.3];
- 1 for the correlogram [4.7.4];
- the correlation coefficient for the cross correlogram [4.7.4].

Variogram surfaces are represented as a pixel map using one of the color palettes found under the **Graph I Palette** menu item. They can be saved as a grid file [6.3], which can then be displayed as pixel maps with the Grid Display program.

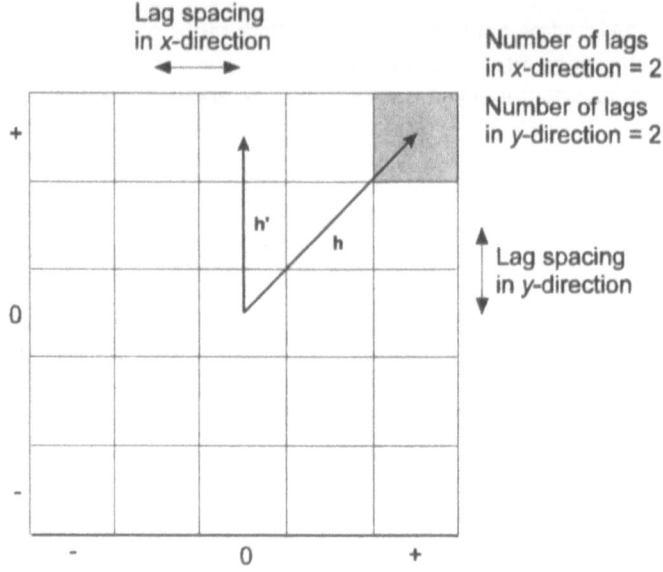

Figure 4.7 Parameters required to calculate a (cross) variogram surface.

A variogram surface can also be displayed in a table format (use **Options** I **Numerical Results !** or **right click**) and saved as a variogram surface file [6.4] for further processing.

The **Graph** I **Cov./Corr.** menu item (function key **F8**) displays the value of the covariance/variance when the estimator is the (cross) variogram [4.7.1] or the (cross) covariance [4.7.3] and displays the correlation coefficient when the estimator is the standardized (cross) variogram [4.7.2] or the correlogram [4.7.4].

4.5 Directional Variogram and Directional Cross Variogram

In Vario2D with PCF a directional (cross) variogram is produced using the **Calculate** I **Directional Variogram...** menu item and constructed with the parameters shown in Figure 4.8.

The direction angle is measured counterclockwise from the east (trigonometric angle). The default lag tolerance is equal to half the lag spacing. Making it greater than the lag spacing will produce a smoothed variogram and entering a negative number will reset it to the default value.

Lag 0 is an extra lag that holds the pairs having a separation distance smaller than the lag tolerance.

Calculating a variogram with the default angular tolerance of 90 is equivalent to averaging variograms calculated in all directions. This type of variogram is referred to as an *omnidirectional variogram*. In Vario2D with PCF, an omnidirectional variogram is calculated with the help of all pairs; the (**x**, **x** + **h**) pair is considered as is the (**x** + **h**, **x**) pair. This ensures that the omnidirectional cross covariance [4.7.3] is computed adequately. Using the maximum bandwidth parameter guarantees that a variogram is truly directional. H-scatterplots for a specific lag are obtained by clicking on the point representing the value of a spatial continuity measure for that lag.

A directional variogram can be displayed in a table format (use **Options | Numerical Results !** or **right click**) and saved into a variogram file [6.5], which can then be used to build a 2D nested model of spatial continuity with the Model program.

The **Graph | Cov./Corr.** menu item (function key **F8**) displays the value of the covariance/variance when the estimator is the (cross) variogram [4.7.1] or the (cross) covariance [4.7.3] and displays the correlation coefficient when the estimator is the standardized (cross) variogram [4.7.2] or the correlogram [4.7.4].

The **Graph | Axis...** menu item gives the user the opportunity to rescale the

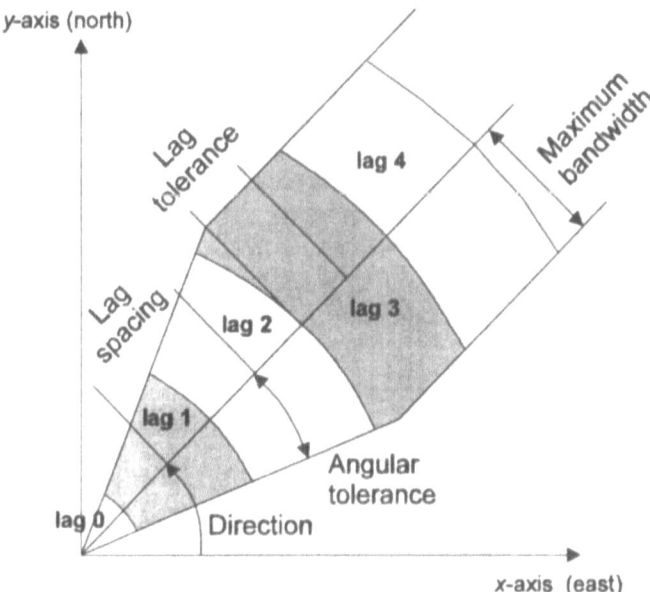

Figure 4.8 Parameters required to calculate a directional (cross) variogram.

variogram plot. This rescaling allows for the visual comparison of several directional variograms.

See also

Model – interactive variogram modeling [Chapter 5].

4.6 Variogram Cloud and Cross Variogram Cloud

In Vario2D with PCF a variogram cloud is produced using the **Calculate |
Variogram Cloud...** menu item and constructed with the parameters shown in Figure 4.9. The direction angle is measured counterclockwise from the east and corresponds to the trigonometrical angle. Using the maximum bandwidth parameter guarantees that a variogram cloud is truly directional.

An omnidirectional variogram cloud can be calculated by setting the angular tolerance to 90. In that case, all $(\mathbf{x}, \mathbf{x} + \mathbf{h})$ pairs as well as their symmetrics, $(\mathbf{x} + \mathbf{h}, \mathbf{x})$, are considered.

A variogram cloud is skewed along the y-axis. Indeed, if the data are a realization of a Gaussian process, then each variogram value [4.7.1] is a

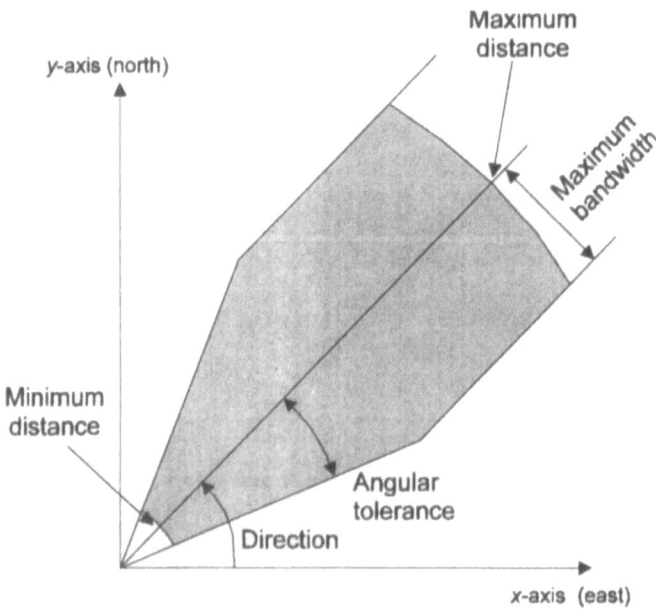

Figure 4.9 Parameters required to calculate (cross) variogram cloud.

realization of a random variable [A.1] proportional to a χ_1^2 variable. Since the experimental variogram can be considered as the moving average of the variogram cloud (Figure 4.1), the pairs having very large variogram values are worthy of particular attention since they contribute greatly to the experimental variogram estimates. If the sample map is on-screen (as an active window or as an icon), pairs selected on a variogram cloud are immediately displayed on the map (use **Ctrl + left click** to select multiple pairs and **left click** to end the selection). The variogram cloud can also help in determining the optimum lag spacing to use when calculating a directional variogram.

A variogram cloud can be calculated in any direction with any angular tolerance and it can be saved in a variogram cloud file [6.6] for further processing (for instance, running a resistant smoother on it or calculating linear forms other than the square difference). Note that a variogram cloud calculated with an angular tolerance close to 90°, a minimum distance of 0, and a maximum distance greater than or equal to the maximum pair distance will display all pairs contained in the pair comparison file [6.2]. Thus, saving a (cross) variogram cloud to a file is a way to extract selected pairs from a pair comparison file.

4.7 Measures of Spatial Continuity Available Within Vario2D with PCF

Several measures of spatial continuity are computed each time a variogram surface or a directional variogram is constructed. The choice for displaying a measure is made in the **Options | Estimator** menu, and the **F1** through **F5** function keys provide a quick way to switch from one measure to another.

All measures of spatial continuity are a statistical summary of an associated h-scatterplot. Some of them – the standardized variogram, the covariance, and the correlogram – include a centering or a standardizing term that depends on the data values located at the origin and at the end of the separation vector **h**. This rescaling by the local mean and/or lag variance makes these *nonergodic* estimators more robust and resistant than the traditional variogram [Isaaks & Srivastava, 1988; Srivastava & Parker, 1989].

4.7.1 Variogram and Cross Variogram

Variogram

The experimental *variogram*, also referred to as the *semivariogram*, for a separation vector **h** is calculated according to the following formula:

$$\gamma(\mathbf{h}) \;=\; \frac{1}{2N(\mathbf{h})} \sum_{i=1}^{N(\mathbf{h})} \left[z(\mathbf{x}_i) - z(\mathbf{x}_i + \mathbf{h}) \right]^2 .$$

When the separation vector **h** is null, this expression always yields 0.
The variogram is symmetric in **h**:

$$\gamma(\mathbf{h}) \;=\; \gamma(-\mathbf{h}).$$

Figure 4.10 shows that the variogram can be considered as the moment of inertia around the diagonal of the associated h-scatterplot.

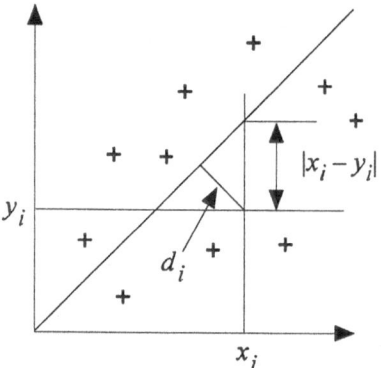

$$I = \sum_i m_i d_i^2 \;=\; \sum_i \frac{1}{N(\mathbf{h})} d_i^2 \;=\; \frac{1}{N(\mathbf{h})} \sum_i \frac{1}{2} (x_i - y_i)^2 \;=\; \frac{1}{2N(\mathbf{h})} \sum_i (x_i - y_i)^2$$

Figure 4.10 The variogram considered as the moment of inertia around the diagonal line of an h-scatterplot.

Cross Variogram

The experimental *cross variogram*, also referred to as the *cross semivariogram*, for a separation vector **h** is calculated according to the following formula:

$$\gamma_{12}(\mathbf{h}) \;=\; \frac{1}{2N(\mathbf{h})} \sum_{i=1}^{N(\mathbf{h})} \left[z_1(\mathbf{x}_i) - z_1(\mathbf{x}_i + \mathbf{h}) \right] \cdot \left[z_2(\mathbf{x}_i) - z_2(\mathbf{x}_i + \mathbf{h}) \right].$$

When the separation vector **h** is null, this expression always yields 0.

The cross variogram is symmetric in **h** and in the variables:

$$\gamma_{12}(\mathbf{h}) = \gamma_{12}(-\mathbf{h}),$$
$$\gamma_{12}(\mathbf{h}) = \gamma_{21}(\mathbf{h}).$$

See also

second-order moments [A.2.2],
second-order stationarity [A.4.2],
intrinsic hypothesis [A.4.3].

4.7.2 Standardized Variogram and Standardized Cross Variogram

Standardized Variogram

The experimental *standardized variogram* for a separation vector **h** is computed according to the following formula:

$$\gamma_S(\mathbf{h}) = \frac{\gamma(\mathbf{h})}{\sigma_{-\mathbf{h}} \cdot \sigma_{+\mathbf{h}}},$$

where $\gamma(\mathbf{h})$ is the variogram for the separation vector **h**,

$$\sigma_{-\mathbf{h}}^2 = \frac{1}{N(\mathbf{h})} \sum_{i=1}^{N(\mathbf{h})} z^2(\mathbf{x}_i) - m_{-\mathbf{h}}^2, \quad m_{-\mathbf{h}} = \frac{1}{N(\mathbf{h})} \sum_{i=1}^{N(\mathbf{h})} z(\mathbf{x}_i),$$

$$\sigma_{+\mathbf{h}}^2 = \frac{1}{N(\mathbf{h})} \sum_{i=1}^{N(\mathbf{h})} z^2(\mathbf{x}_i + \mathbf{h}) - m_{+\mathbf{h}}^2, \quad m_{+\mathbf{h}} = \frac{1}{N(\mathbf{h})} \sum_{i=1}^{N(\mathbf{h})} z(\mathbf{x}_i + \mathbf{h}).$$

When the separation vector **h** is null, this expression always yields 0.

This measure rescales the variogram by the lag variance. It has properties similar to those of the *general relative variogram* [Srivastava & Parker, 1989; Isaaks & Srivastava, 1989] and is able to take into account the influence of a locally changing variability of the data.

The standardized variogram is symmetric in **h**:

$$\gamma_S(\mathbf{h}) = \gamma_S(-\mathbf{h}).$$

For an omnidirectional measure, that is, a directional variogram with an angular tolerance of 90°, the standardized variogram and the correlogram [4.7.4] are linked according to the following equation:

$$\gamma_S(h) = 1 - \rho(h).$$

Standardized Cross Variogram

The experimental *standardized cross variogram* for a separation vector **h** is calculated according to the following formula:

$$\gamma_{S_{12}}(h) = \frac{\gamma_{12}(h)}{\sqrt{\sigma_{z_{1-h}} \cdot \sigma_{z_{1+h}}} \cdot \sqrt{\sigma_{z_{2-h}} \cdot \sigma_{z_{2+h}}}},$$

where $\gamma_{12}(h)$ is the cross variogram for the separation vector **h**,

$$\sigma^2_{z_{1-h}} = \frac{1}{N(h)} \sum_{i=1}^{N(h)} z_1^2(x_i) - m^2_{z_{1-h}}, \quad m_{z_{1-h}} = \frac{1}{N(h)} \sum_{i=1}^{N(h)} z_1(x_i),$$

$$\sigma^2_{z_{1+h}} = \frac{1}{N(h)} \sum_{i=1}^{N(h)} z_1^2(x_i + h) - m^2_{z_{1+h}}, \quad m_{z_{1+h}} = \frac{1}{N(h)} \sum_{i=1}^{N(h)} z_1(x_i + h),$$

$$\sigma^2_{z_{2-h}} = \frac{1}{N(h)} \sum_{i=1}^{N(h)} z_2^2(x_i) - m^2_{z_{2-h}}, \quad m_{z_{2-h}} = \frac{1}{N(h)} \sum_{i=1}^{N(h)} z_2(x_i),$$

$$\sigma^2_{z_{2+h}} = \frac{1}{N(h)} \sum_{i=1}^{N(h)} z_2^2(x_i + h) - m^2_{z_{2+h}}, \quad m_{z_{2+h}} = \frac{1}{N(h)} \sum_{i=1}^{N(h)} z_2(x_i + h).$$

When the separation vector **h** is null, this expression yields 0.

This measure rescales the cross variogram by the lag variance of both variables.

The standardized cross variogram is symmetric in **h** and in the variables:

$$\gamma_{S_{12}}(h) = \gamma_{S_{12}}(-h),$$

$$\gamma_{S_{12}}(h) = \gamma_{S_{21}}(h).$$

See also

second-order stationarity [A.4.2].

4.7.3 Covariance and Cross Covariance

Covariance

The experimental *covariance* for a separation vector **h** is calculated according to the following formula:

$$C(\mathbf{h}) = \frac{1}{N(\mathbf{h})} \sum_{i=1}^{N(\mathbf{h})} z(\mathbf{x}_i) \cdot z(\mathbf{x}_i + \mathbf{h}) - m_{-\mathbf{h}} \cdot m_{+\mathbf{h}},$$

where

$$m_{-\mathbf{h}} = \frac{1}{N(\mathbf{h})} \sum_{i=1}^{N(\mathbf{h})} z(\mathbf{x}_i), \quad \text{and} \quad m_{+\mathbf{h}} = \frac{1}{N(\mathbf{h})} \sum_{i=1}^{N(\mathbf{h})} z(\mathbf{x}_i + \mathbf{h}).$$

When the separation vector **h** is null, this expression yields the statistical variance.

Since the centering term is the product of two different lags means, this measure of spatial continuity is also referred to as the *nonergodic covariance* [Isaaks & Srivastava, 1988].

The covariance is symmetric in **h**:

$$C(\mathbf{h}) = C(-\mathbf{h}).$$

Cross Covariance

The experimental *cross covariance* for a separation vector **h** is calculated according to the following formula:

$$C_{12}(\mathbf{h}) = \frac{1}{N(\mathbf{h})} \sum_{i=1}^{N(\mathbf{h})} z_1(\mathbf{x}_i) \cdot z_2(\mathbf{x}_i + \mathbf{h}) - m_{z_{1-\mathbf{h}}} \cdot m_{z_{2+\mathbf{h}}},$$

where

$$m_{z_{1-\mathbf{h}}} = \frac{1}{N(\mathbf{h})} \sum_{i=1}^{N(\mathbf{h})} z_1(\mathbf{x}_i), \quad \text{and} \quad m_{z_{2+\mathbf{h}}} = \frac{1}{N(\mathbf{h})} \sum_{i=1}^{N(\mathbf{h})} z_2(\mathbf{x}_i + \mathbf{h}).$$

When the separation vector **h** is null, this expression yields the statistical covariance.

The cross covariance is not the same if it is computed in the opposite direction. Reversing the x- and y-values on a cross h-scatterplot entails switching not only the direction of the separation vector $\mathbf{h}$ but also the order of the variables. The following relationships hold:

$$C_{12}(\mathbf{h}) \neq C_{12}(-\mathbf{h}),$$
$$C_{12}(\mathbf{h}) = C_{21}(-\mathbf{h}).$$

See also

second-order moments [A.2.2],
second-order stationarity [A.4.2].

4.7.4 Correlogram and Cross Correlogram

Correlogram

The experimental *correlogram* for a separation vector $\mathbf{h}$ is calculated according to the following formula:

$$\rho(\mathbf{h}) = \frac{C(\mathbf{h})}{\sigma_{-\mathbf{h}} \cdot \sigma_{+\mathbf{h}}},$$

where $C(\mathbf{h})$ is the covariance for the separation vector $\mathbf{h}$,

$$\sigma^2_{-\mathbf{h}} = \frac{1}{N(\mathbf{h})} \sum_{i=1}^{N(\mathbf{h})} z^2(x_i) - m^2_{-\mathbf{h}}, \quad m_{-\mathbf{h}} = \frac{1}{N(\mathbf{h})} \sum_{i=1}^{N(\mathbf{h})} z(x_i),$$

$$\sigma^2_{+\mathbf{h}} = \frac{1}{N(\mathbf{h})} \sum_{i=1}^{N(\mathbf{h})} z^2(x_i + \mathbf{h}) - m^2_{+\mathbf{h}}, \quad m_{+\mathbf{h}} = \frac{1}{N(\mathbf{h})} \sum_{i=1}^{N(\mathbf{h})} z(x_i + \mathbf{h}).$$

When the separation vector $\mathbf{h}$ is null, this expression yields 1. Note also that the value of this measure of spatial continuity is always between 0 and 1.

Since the rescaling term is the product of two different lags standard deviations, this measure of spatial continuity is also refered to as the *nonergodic correlogram* [Isaaks & Srivastava, 1988].

The correlogram is symmetric in $\mathbf{h}$:

$$\rho(\mathbf{h}) = \rho(-\mathbf{h}).$$

For an omnidirectional measure, that is, a directional variogram with an angular tolerance of 90°, the standardized variogram and the correlogram are linked according to the following equation:

$$\rho(h) = 1 - \gamma_S(h).$$

Cross Correlogram

The experimental *cross correlogram* for a separation vector **h** is calculated according to the following formula:

$$\rho_{12}(h) = \frac{C_{12}(h)}{\sigma_{z_{1-h}} \cdot \sigma_{z_{2+h}}},$$

where $C_{12}(h)$ is the cross covariance for the separation vector **h**,

$$\sigma^2_{z_{1-h}} = \frac{1}{N(h)} \sum_{i=1}^{N(h)} z_1^2(x_i) - m^2_{z_{1-h}}, \quad m_{z_{1-h}} = \frac{1}{N(h)} \sum_{i=1}^{N(h)} z_1(x_i),$$

$$\sigma^2_{z_{2+h}} = \frac{1}{N(h)} \sum_{i=1}^{N(h)} z_2^2(x_i + h) - m^2_{z_{2+h}}, \quad m_{z_{2+h}} = \frac{1}{N(h)} \sum_{i=1}^{N(h)} z_2(x_i + h).$$

When the separation vector **h** is null, this expression yields the statistical correlation coefficient. Note also that the value of this measure of spatial continuity is always between −1 and +1.

The cross correlogram is not the same if it is computed in the opposite direction. Reversing the *x*- and *y*-values on a cross h-scatterplot entails switching not only the direction of the separation vector **h** but also the order of the variables. The following relationships hold:

$$\rho_{12}(h) \neq \rho_{12}(-h),$$
$$\rho_{12}(h) = \rho_{21}(-h).$$

See also

second-order stationarity [A.4.2].

4.7.5 Madogram and Cross Madogram

Madogram

The experimental *madogram*, or mean absolute deviation, for a separation vector **h** is calculated according to the following formula:

$$M(\mathbf{h}) = \frac{1}{2N(\mathbf{h})} \sum_{i=1}^{N(\mathbf{h})} \left| z(\mathbf{x}_i) - z(\mathbf{x}_i + \mathbf{h}) \right|.$$

When the separation vector **h** is null, this expression yields 0.

The madogram is symmetric in **h**:

$$M(\mathbf{h}) = M(-\mathbf{h}).$$

If the random function representing the phenomenon is multi-Gaussian [A.6], then the following relation is true for all separation vectors **h**:

$$\frac{\sqrt{\gamma(\mathbf{h})}}{M(\mathbf{h})} = \sqrt{\pi}.$$

Cross Madogram

The experimental *cross madogram* for a separation vector **h** is calculated according to the following formula:

$$M_{12}(\mathbf{h}) = \frac{1}{2N(\mathbf{h})} \sum_{i=1}^{N(\mathbf{h})} \sqrt{\left| z_1(\mathbf{x}_i) - z_1(\mathbf{x}_i + \mathbf{h}) \right|} \cdot \sqrt{\left| z_2(\mathbf{x}_i) - z_2(\mathbf{x}_i + \mathbf{h}) \right|}.$$

When the separation vector **h** is null, this expression is set to 0.

The cross madogram is symmetric in **h** and in the variables:

$$M_{12}(\mathbf{h}) = M_{12}(-\mathbf{h}),$$
$$M_{12}(\mathbf{h}) = M_{21}(\mathbf{h}).$$

Further Reading

Armstrong, M., "Common Problems Seen in Variograms," *Mathematical Geology*, Vol. 16, No. 3, pp. 305–313, 1984.

Bradley, R. & Haslett, J., "Interactive Graphics for the Exploratory Analysis of Spatial Data – The Interactive Variogram Cloud," in Proc. of the Second CODATA Conference on Geomathematics and Geostatistics, P. A. Dowd & J. J. Royer (eds.), *Sci. de la Terre, Sér. Inf.*, Nancy, No. 31, pp. 373–386, 1992.

Carr, J. R., Bailey, R. E. & Deng, E. D., "Use of Indicator Variograms for an Enhanced Spatial Analysis," *Mathematical Geology*, Vol. 17, No. 8, pp. 797–811, 1985.

Chauvet, P., "The Variogram Cloud," Internal Note N-725, Centre de Géostatistique, Fontainebleau, France, 1982.

David, M., "Geostatistical Ore Reserve Estimation," Elsevier, New York, 1977.

Isaaks, E. H. & Srivastava, R. M., "Spatial Continuity Measures for Probabilistic and Deterministic Geostatistics," *Mathematical Geology*, Vol. 20, No. 4, pp. 313–341, 1988.

Isaaks, E. H. & Srivastava, R. M., "An Introduction to Applied Geostatistics," Oxford University Press, New York, NY, 1989.

Journel, A. G., "Fundamentals of Geostatistics in Five Lessons," *Short Course in Geology*, Vol. 8, American Geophysical Union, Washington, DC, 1989.

Rendu, J., "Kriging for Ore Valuation and Mine Planning," *Engineering and Mining Journal*, Vol. 181, No. 1, pp. 114–120, 1980.

Schofield, N., "Ore Reserve Estimation at the Enterprise Gold Mine, Pine Creek, Northern Territory, Australia. Part 1: Structural and Variogram Analysis," *CIM Bulletin*, Vol. 81, No. 909, pp. 56–61, 1988.

Srivastava, R. M. & Parker, H. M., "Robust Measures of Spatial Continuity in Geostatistics," in *Geostatistics*, M. Armstrong (ed.), Proc. of the Third International Geostatistics Congress, September 5–9, 1988, Avignon, France, D. Reidel, Dordrecht, Holland, pp. 295–308, 1989.

Verly, G., "Estimation of Spatial Point and Block Distributions: The Multi-Gaussian Model," Ph.D. thesis, Stanford University, Stanford, CA, 1984.

5
Model – Interactive Variogram Modeling

5.1 Overview

The Model program constructs a 2D nested model with the help of experimental (cross) variogram(s) produced by Vario2D with PCF [4.5]. It deals with 2D anisotropic modeling of one or two variables but does not provide any facility for constructing a global coherent model of coregionalization.

Directional variograms are read from a variogram file [6.5]. Scroll bars can change the parameters of the 2D nested model. Each time a parameter is modified, cross sections through the 2D model are redrawn along with the directional variogram(s) used to fit the model.

A 2D nested model can be fitted against any of the measures of spatial continuity available in Vario2D with PCF [4.7].

An indicative goodness of fit (IGF) is computed every time the model changes. The best IGF is stored in memory and can be recalled at any time. A user's model can also be stored in memory.

The 2D nested model of spatial continuity can be saved as a grid file [6.3]. This file can then be used to produce a variogram surface representation of the 2D nested model.

5.2 Variogram Models

Variogram models cannot be just any function. They must satisfy a positive-definite condition [A.5]. One way to satisfy the positive-definite condition is to use only a few functions that are known to be positive definite. Although this may seem restrictive, those functions can be linearly combined to form new functions that are also valid. The resulting model, referred to as a *nested model*, can be expressed as

$$\gamma(\mathbf{h}) = \sum_{i=1}^{n} w_i \gamma_i(\mathbf{h}),$$

where $\gamma_i(\mathbf{h})$ is a function known to be positive definite, and w_i is an associated weight. The sum of those weights should be equal to the (co)variance of the modeled random function [A.2.2].

The following conditions ensure that the nested model is a valid function:

- all models found in the weighted linear combination, the so-called *nested structures*, are valid;
- all weights found in the weighted linear combination are greater than or equal to 0; if some weights are negative, the nested model may or may not be valid.

The positive-definite variogram models presented in this section are considered to be the *basic* models. They are simple, isotropic models, that depend only on the magnitude, |h|, of the separation vector **h**. We will see later in this chapter how they can be transformed into anisotropic functions that depend not only on the magnitude of the separation vector **h**, but also on its direction [5.3.2]. The basic models can be conveniently divided into two groups:

- those that reach a plateau; and
- those that do not.

Variogram models of the first type are the so called *transition* models. They are used to model second-order stationary, random functions [A.4.2]. The plateau they reach is called the *sill*, and the distance at which they reach this plateau is called the *range*. Some of the transition models reach their sill asymptotically. For such a model, the range is defined to be the distance at which 95% of the sill is reached.

Variogram models of the second type do not reach a sill; instead, they continue to increase as the distance increases. Such models are used to model intrinsic random functions [A.4.3].

5.2.1 Nugget Effect Model

The *nugget effect* model is used to model a discontinuity at the origin. While the (cross) variogram [4.7.1] value for |h| = 0 is strictly 0, the (cross) variogram value at a very small separation distance may be significantly different from 0, giving rise to a discontinuity. Such a discontinuity can be modeled using a discontinuous, positive definite transition model that is 0 when |h| is equal to 0 and that is the value of the discontinuity otherwise. This is the nugget effect, model and its equation is given by:

$$\gamma(|\mathbf{h}|) = c = \begin{cases} 0 & \text{if } |\mathbf{h}| = 0, \\ c & \text{otherwise,} \end{cases}$$

where c is the discontinuity at the origin.

While other models of variograms can be anisotropic, the Model program requires the nugget effect model to be isotropic.

If second-order stationarity [A.4.2] is assumed, the nugget effect can be expressed as a covariance, a correlogram, or a standardized variogram using the following relationships:

$$C(\mathbf{h}) = C(0) - \gamma(\mathbf{h}),$$
$$\rho(\mathbf{h}) = \rho(0) - \gamma_S(\mathbf{h}).$$

5.2.2 Spherical Model

The *spherical model* is probably the most commonly used transition model. Its equation is given by

$$\gamma(|\mathbf{h}|) = c \cdot \mathrm{Sph}_a(|\mathbf{h}|) = \begin{cases} c \cdot \left[1.5 \dfrac{|\mathbf{h}|}{a} - 0.5 \left(\dfrac{|\mathbf{h}|}{a} \right)^3 \right] & \text{if } |\mathbf{h}| \le a, \\ c & \text{otherwise,} \end{cases}$$

where a is the range and c is the (co)variance contribution or sill value.

This model has linear behavior at the origin and reaches the sill at distance

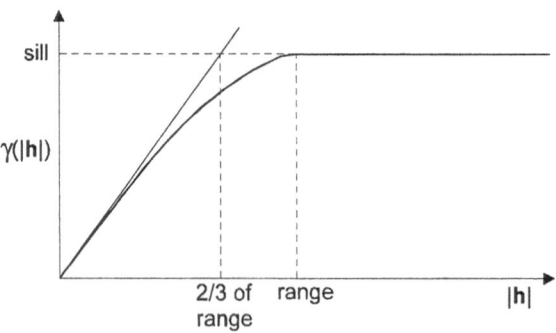

Figure 5.1 The spherical model.

a. In fitting this model, it is useful to remember that the tangent at the origin reaches the sill at two-thirds of the range (Figure 5.1).

If second-order stationarity [A.4.2] is assumed, the spherical model can be expressed as a covariance, a correlogram, or a standardized variogram using the following relationships:

$$C(\mathbf{h}) = C(0) - \gamma(\mathbf{h}),$$
$$\rho(\mathbf{h}) = \rho(0) - \gamma_S(\mathbf{h}).$$

5.2.3 Exponential Model

The *exponential model* is another commonly used transition model. Its equation is given by

$$\gamma(|\mathbf{h}|) = c \cdot \mathrm{Exp}_a(|\mathbf{h}|) = c \cdot \left[1 - e^{-\frac{3|\mathbf{h}|}{a}} \right],$$

where c is the (co)variance contribution or sill value and a is the practical range, that is, the distance at which the variogram value is 95% of the sill.

This model reaches its sill asymptotically and has linear behavior at the origin. In fitting this model, it is useful to remember that the tangent at the origin reaches the sill at one-third of the practical range (Figure 5.2).

If second-order stationarity [A.4.2] is assumed, the exponential model can be expressed as a covariance, a correlogram, or a standardized variogram using

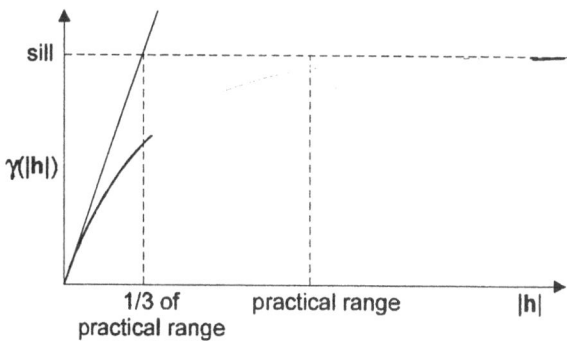

Figure 5.2 The exponential model.

the following relationships:

$$C(\mathbf{h}) = C(0) - \gamma(\mathbf{h}),$$
$$\rho(\mathbf{h}) = \rho(0) - \gamma_S(\mathbf{h}).$$

5.2.4 Gaussian Model

The *Gaussian model* is a transition model used to model extremely continuous phenomena. Its equation is given by

$$\gamma(|\mathbf{h}|) = c \cdot \mathrm{Gauss}_a(|\mathbf{h}|) = c \cdot \left[1 - e^{-\frac{3|\mathbf{h}|^2}{a^2}} \right],$$

where c is the positive variance contribution or sill value and a is the practical range, namely, the distance at which the variogram value is 95% of the sill.

This model reaches its sill asymptotically and has parabolic behavior at the origin. It is the only basic transition model whose shape has an inflection point (Figure 5.3).

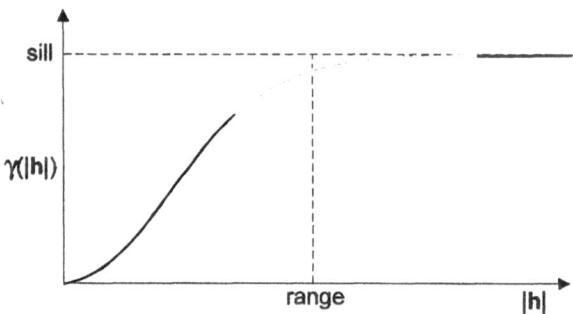

Figure 5.3 The Gaussian model.

If second-order stationarity [A.4.2] is assumed, the Gaussian model can be expressed as a covariance, a correlogram, or a standardized variogram using the following relationships:

$$C(\mathbf{h}) = C(0) - \gamma(\mathbf{h}),$$
$$\rho(\mathbf{h}) = \rho(0) - \gamma_S(\mathbf{h}).$$

Matrix instability problems in kriging are often encountered when using a Gaussian model without a nugget effect [Posa, 1988]. A solution to this problem is to add a small nugget effect to the Gaussian model.

5.2.5 Power Model

The *power model* is not a transition model since it does not reach a sill but increases with the magnitude of **h**. Its equation is given by

$$\gamma(|\mathbf{h}|) = c \cdot \mathrm{Pow}_a(|\mathbf{h}|) = c \cdot |\mathbf{h}|^a,$$

where c is the positive variance contribution and a is a power between 0 and 2 $(0 < a < 2)$.

This model is of particular interest since it can show a whole range of behavior at the origin depending on the value of a (Figure 5.4). When the a parameter is equal 1, this model is also referred to as a *linear variogram*.

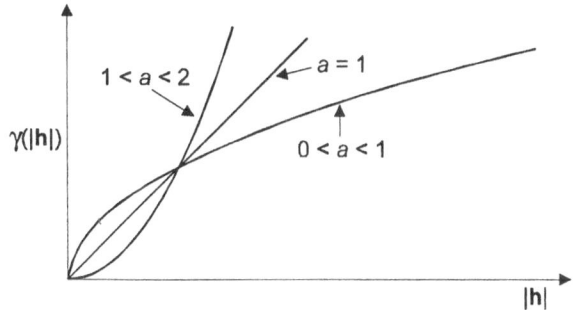

Figure 5.4 The power model.

This model is valid [A.5] only for random functions that are intrinsic [A.4.3]. It should not be used for second-order stationary, random functions [A.4.2].

5.3 Interactive Construction of a 2D Nested Model

After a variogram file [6.5] has been read into memory (use the **File | Open...** menu item), selecting the **Calculate | Model...** menu item displays a list of all directional (cross) variograms that have just been read from the file. One or several directional (cross) variograms must be selected from this list (use **left click** or **Ctrl + left click**). Note that all selected directional (cross) variograms must relate to the same variable(s). The a priori (co)variance and correlation coefficient used as default for fitting the 2D nested model against the covariance and the correlogram are the (co)variance and the correlation coefficient of the first selected directional (cross) variogram. Once the **OK** button is pressed, two windows appear:

- the *modeling window*, which contains various controls and scroll bars;
- the *plot window*, which contains a plot of the selected directional (cross) variograms.

The 2D model parameters are modified in the modeling window. Each time a parameter is changed, the 2D nested model is recalculated along with its indicative goodness of fit and cross sections through the 2D model are eventually plotted, along with the corresponding experimental variograms, on the plot window. Since these two windows are closely related, closing one will close the other, iconizing one will iconize the other, and restoring one will restore the other.

Since Model is a multiple-document interface (MDI) application, both windows can be reduced to icons and several (cross) variograms can be modeled during the same session, provided that all necessary directional (cross) variograms are stored in the same variogram file (use the **Calculate I Model...** menu item).

5.3.1 The Modeling Window

Figure 5.5 shows the modeling window with its scroll bars and buttons. Parameters of the model can be changed with scroll bars or directly by entering values with the keyboard. Nugget and sill scroll bars have default maximum values, which are equal to the rounded a priori covariance/correlation coefficient. These two values can be modified using the **A priori Cov./Corr. >>** button. Range scroll bars have default maximum values, which are equal to the rounded maximum distance found in the experimental variograms used for modeling. Remember that a **left click on the left/right button of a scroll bar** decreases/increases the corresponding parameter by 1/100 of the scroll bar range. A **left click** on the left/right of the scroll bar cursor decreases/increases the corresponding parameter by 1/10 of the scroll bar range.

The 2D nested model built in this window is the sum of an isotropic nugget effect and one, two, or three nested structures, which are most often anisotropic in 2D.

When the modeling window is the active window, the **Estimator** menu item is enabled and various measures of spatial continuity can be chosen for fitting the 2D nested model [4.7]. When changing measures, the nested model already constructed is translated in terms compatible with the new chosen measure.

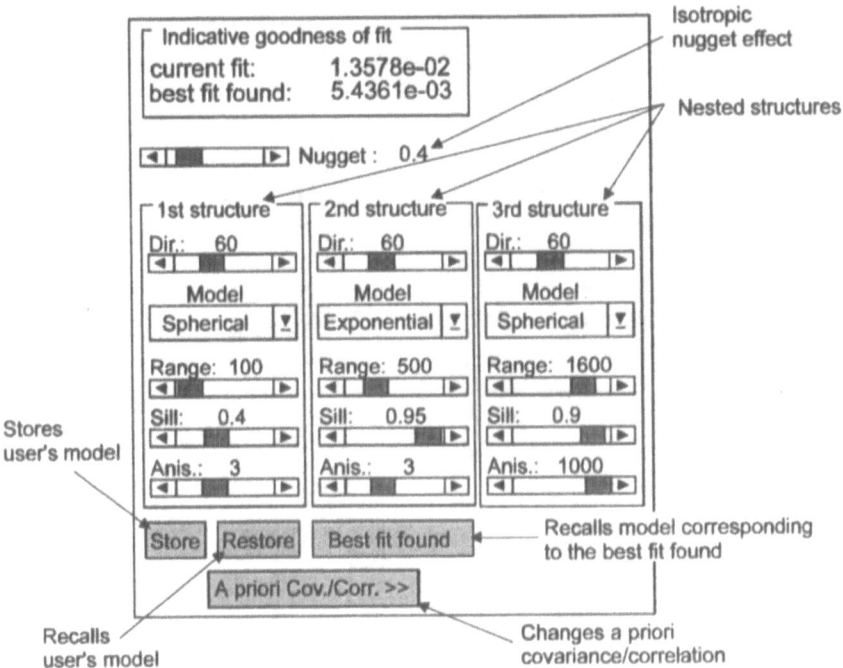

Figure 5.5 The modeling window.

Using the **File | Save as...** menu item when the modeling window is the active window makes it possible to save the current nested model in a model file [6.7]. This file can then be used to specify the 2D model parameters in a (co)kriging or a (co)simulation program.

5.3.2 Modeling a Nested Structure

The parameters found in the modeling window in a box corresponding to a nested structure provide a complete description of an anisotropic 2D basic model whose range changes with direction (Figure 5.6). This behavior is referred to as *geometric anisotropy*. In order to adjust the 2D anisotropic basic model, directional (cross) variograms should be displayed on the plot window for directions of maximum and minimum continuity as well as for some intermediate directions.

A 2D anisotropic basic model can be described with an isotropic basic model applied to to the magnitude of a reduced vector:

$$\gamma(\mathbf{h}) = \gamma\left(\left|\mathbf{h}_{\text{reduced}}\right|\right)$$

Calculating a reduced vector is equivalent to defining a new coordinate system in which the ellipse representing the 2D anisotropic model is transformed into a circle. Calculation of the reduced distance is performed according to the following procedure:

1. The first step is a rotation of the axes by the angle corresponding to the **Dir.** parameter shown in Figure 5.6. This operation makes the new axes parallel to the main axes of the ellipse:

$$\begin{bmatrix} x_1 \\ y_1 \end{bmatrix} = [R_{Dir.}]\begin{bmatrix} h_x \\ h_y \end{bmatrix}, \quad \text{with} \quad [R_{Dir.}] = \begin{bmatrix} \cos(Dir.) & \sin(Dir.) \\ -\sin(Dir.) & \cos(Dir.) \end{bmatrix}.$$

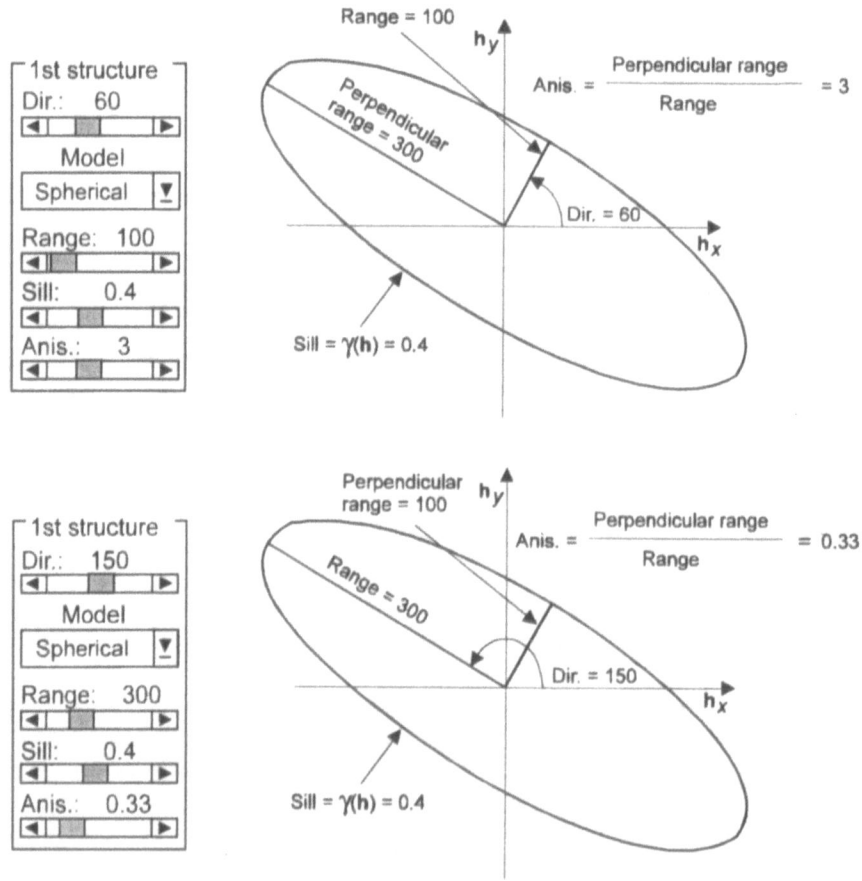

Figure 5.6 Two equivalent ways of describing an anisotropic basic model in 2D.

2. The second step is the transformation of the ellipse into a circle of radius equal to the range of the ellipse along one of its two main axes. This radius is given by the **Range** parameter shown in Figure 5.6. This is equivalent to squeezing or extending the ellipse along one of its two axes by an amount given by the **Anis.** parameter:

$$\begin{bmatrix} x \\ y \end{bmatrix} = [D_{Anis}] \begin{bmatrix} x_1 \\ y_1 \end{bmatrix}, \quad \text{with} \quad [D_{Anis}] = \begin{bmatrix} 1 & 0 \\ 0 & \dfrac{1}{Anis.} \end{bmatrix}.$$

3. Finally, the 2D anisotropic basic model is given by

$$\gamma(|\mathbf{h}_{reduced}|) = \gamma\left(\sqrt{x^2 + y^2}\right).$$

Isotropy is the special case where the **Anis.** parameter is 1 (the default value). When the **Anis.** parameter is set to a large value (e.g. 1000), the contribution of the basic model in the direction perpendicular to the **Dir.** parameter is insignificant. Such a case is referred to as a *zonal anisotropy*, where the sill of the 2D model changes with direction (Figure 5.7).

Note that the anisotropy of the power model is handled in a correct manner; an anisotropic distance is calculated and the power *a* is left unchanged.

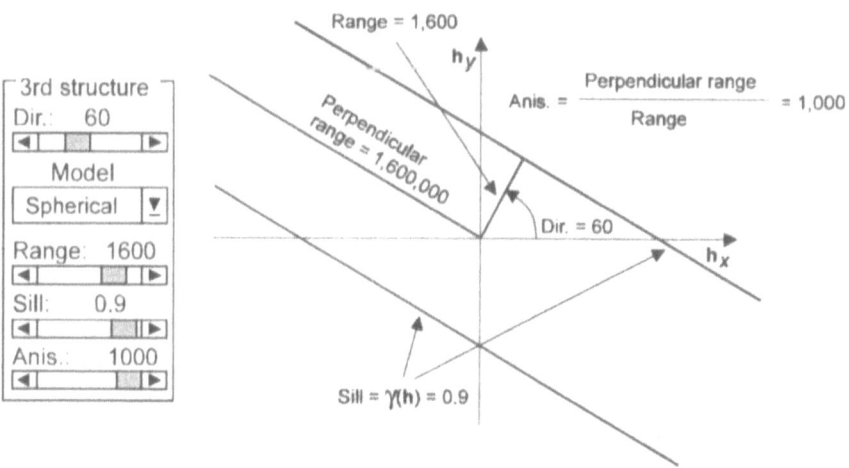

Figure 5.7 Modeling zonal anisotropy.

5.3.3 Indicative Goodness of Fit (IGF)

An indicative goodness of fit is calculated every time the 2D nested model is modified. It gives a measure of how well the cross sections through the 2D nested model adjust the experimental (cross) variograms displayed on the plot window [5.3.4]. It is calculated with the following equation:

$$\text{IGF} = \frac{1}{N}\sum_{k=1}^{N}\sum_{i=0}^{n(k)}\frac{P(i)}{\sum_{j=0}^{n(k)}P(j)}\cdot\frac{D(k)}{d(i)}\cdot\left[\frac{\gamma(i)-\hat{\gamma}(i)}{\sigma^2}\right]^2,$$

where N is the number of directional variograms on the plot window,

$n(k)$ is the number of lags relative to variogram k,

$D(k)$ is the maximum distance relative to variogram k,

$P(i)$ is the number of pairs for lag i of variogram k,

$d(i)$ is the mean pair distance for lag i of variogram k,

$\gamma(i)$ is the experimental measure of spatial continuity for lag i,

$\hat{\gamma}(i)$ is the modeled measure of spatial continuity for $d(i)$,

σ^2 is the (co)variance of the data for the (cross) variogram and the (cross) covariance, the maximum absolute experimental value of all measures for the (cross) madogram, the correlation coefficient for the (cross) correlogram and the (cross) standardized variogram.

The IGF is a number without units; a value close to zero indicates a good fit. Since it is a standardized measure of fit, its value is comparable from one modeling session to another, allowing the user to numerically check how well his models fit the experimental measures.

This indicative goodness of fit has been included as an attempt to quantify the traditionnal visual fit. However, it should be clear that the 2D nested model that can be best adjusted to the experimental variograms might not be the most appropriate one.. It is the user's responsablity to check that the 2D nested model also includes all prior information available for the phenomenon whose spatial continuity is being modeled.

5.3.4 The Plot Window

The plot window displays experimental (cross) variograms along with cross sections through the 2D nested model. When a model parameter is changed in the modeling window, the cross sections through the 2D model are updated on the plot window (Figure 5.8).

All directional (cross) variograms are represented with the same scale which can be modified using the **Graph I Axis...** menu item. Using the **File I**

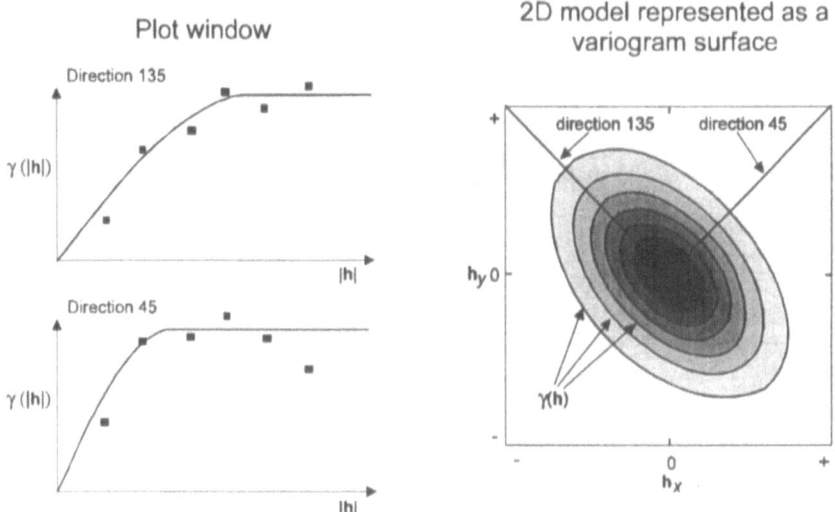

Figure 5.8 A 2D general model viewed along specific directions and its synthetic representation as a variogram surface.

Save as... menu item when the plot window is the active window makes it possible for the 2D nested model to be saved as a grid file [6.3]. The file can then be used to provide a variogram surface representation of the 2D nested model. Comparing the experimental variogram surface and its modeled counterpart with the Grid Display program is an effective way to see what part of the experimental measure has been incorporated in the 2D nested model.

5.4 The Linear Model of Coregionalization

When dealing with several regionalized variables [A.1], each variable is characterized by its own experimental variogram and each pair of variables by its own experimental cross variogram. The functions used to model those variograms and cross variograms must be chosen so that the variance of any possible linear combination of these variables is always positive. The linear model of coregionalization is the most commonly used method for choosing such a set of functions and will be described in the following paragraphs.

We first assume that the K second-order stationary, [A.4.2] random functions $\{ Z_k(\mathbf{x}),\ k = 1, \ldots, K \}$ representing the studied regionalized variables can be defined as linear combinations of N orthogonal (i.e., their cross covariance is always 0) and second-order stationary, random functions $\{ Y_i(\mathbf{x}),\ i = 1, \ldots, N \}$

$$Z_k(\mathbf{x}) = \sum_{i=1}^{N} a_{ki} Y_i(\mathbf{x}).$$

Since the direct covariance [A.2.2] of Y_n is $K_n(\mathbf{h})$, the cross covariance between two random functions $Z_k(\mathbf{x})$ and $Z_{k'}(\mathbf{x})$ is given by

$$C_{kk'}(\mathbf{h}) = \sum_{i=1}^{N} a_{ki} a_{k'i} K_i(\mathbf{h}) = \sum_{i=1}^{N} b_{kk'}^{i} K_i(\mathbf{h}),$$

with

$$b_{kk'}^{i} = a_{ki} a_{k'i}, \quad \forall i = 1 \text{ to } N.$$

The matrix of cross covariances is written as

$$\begin{bmatrix} C_{11} & \cdots & C_{1K} \\ \vdots & \ddots & \vdots \\ C_{K1} & \cdots & C_{KK} \end{bmatrix}$$

$$= \begin{bmatrix} b_{11}^1 & \cdots & b_{1K}^1 \\ \vdots & \ddots & \vdots \\ b_{K1}^1 & \cdots & b_{KK}^1 \end{bmatrix} \cdot \begin{bmatrix} K_1 & \cdots & 0 \\ \vdots & \ddots & \vdots \\ 0 & \cdots & K_1 \end{bmatrix} + \ldots + \begin{bmatrix} b_{11}^N & \cdots & b_{1K}^N \\ \vdots & \ddots & \vdots \\ b_{K1}^N & \cdots & b_{KK}^N \end{bmatrix} \cdot \begin{bmatrix} K_N & \cdots & 0 \\ \vdots & \ddots & \vdots \\ 0 & \cdots & K_N \end{bmatrix}$$

and is *positive-definite* when all matrices $[b_{kk'}^{i}]$ are positive-definite, that is, all those matrices have real and positive eigenvalues. This positive-definite property ensures that the variance of any possible linear combinations of the $Z_k(\mathbf{x})$ is always positive.

The linear model of coregionalization can now be defined with two properties. It is a model where
- all direct and cross covariances are derived from linear combinations of N basic direct covariances $\{ K_i(\mathbf{h}), \quad i = 1, \ldots, N \}$, i.e.,

$$C_{kk'}(\mathbf{h}) = \sum_{i=1}^{N} b_{kk'}^{i} K_i(\mathbf{h}) \quad \text{with} \quad b_{kk'}^{i} = b_{k'k}^{i}, \quad \forall i;$$

this condition can also be written in terms of direct and cross semivariograms:

$$\gamma_{kk'}(\mathbf{h}) = \sum_{i=1}^{N} b_{kk'}^i \gamma_i(\mathbf{h}) \quad \text{with} \quad b_{kk'}^i = b_{k'k}^i, \quad \forall\, i;$$

- for a fixed index i, the coefficient matrix $[\, b_{kk'}^i \,]$ is positive-definite.

Since the coefficient matrices are symmetric, they are positive-definite when their determinant and their diagonal elements are positive or zero. For a matrix of dimension K, the following requirements should be met:

$$b_{11} \geq 0, \quad \begin{vmatrix} b_{11} & b_{12} \\ b_{21} & b_{22} \end{vmatrix} \geq 0, \dots, \quad \begin{vmatrix} b_{11} & \cdots & b_{1K} \\ \vdots & \ddots & \vdots \\ b_{K1} & \cdots & b_{KK} \end{vmatrix} \geq 0.$$

For $K = 2$, verification of the positive-definite condition amounts to checking two inequalities for all i:

$$b_{11}^i \geq 0; \quad \left| b_{12}^i \right| = \left| b_{21}^i \right| \leq \sqrt{b_{11}^i \cdot b_{22}^i}\,.$$

Those requirements on the coefficient matrices imply that a nested structure modeled on a cross variogram must also appear on the two direct variograms.

Model is used to adjust experimental variograms and cross variograms with a 2D nested model. In other words it determines, for a given k and k', the $b_{kk'}^i$, with i ranging from 1 to the number of nested structures. However, it does not provide any facility for constructing a linear model of coregionalization, that is, checking that for a given i the $b_{kk'}^i$, with k and k' spanning all variables, form a positive-definite matrix. It is the user's responsibility to check that the models fitted against the direct and cross variograms are compatible with the linear model of coregionalization. Note that Goulard & Voltz (1992) propose an iterative procedure for building such a set of positive-definite matrices.

Further Reading

Barnes, R. J., "The Variogram Sill and the Sample Variance," *Mathematical Geology*, Vol. 23, No. 4, pp. 673–678, 1991.

Goulard, M. & Voltz, M., "Linear Coregionalization Model: Tools for Estimation and Choice of Cross-Variogram Matrix," *Mathematical Geology*, Vol. 24, No. 3, pp. 269–286, 1992.

Isaaks, E. H. & Srivastava, R. M., "An Introduction to Applied Geostatistics," Oxford University Press, New York, NY, 1989.

Journel, A. G. & Huijbregts, C. J., "Mining Geostatistics," Academic Press, London, 1978.

Myers, D. E., "Pseudo-Cross Variograms, Positive Definiteness and Cokriging," *Mathematical Geology*, Vol. 23, pp. 805–816, 1991.

Posa, D., "Conditioning of the Stationary Kriging Matrices for Some Well-Known Covariance Models," *Mathematical Geology*, Vol. 21, No. 7, pp. 755–765, 1988.

6
Files Used Within VARIOWIN 2.2

Within VARIOWIN, all steps of a variographic study can be recorded into files. This section describes the format, the default extension, and the usage of those files. A user who wishes to modify a file created by one of the programs included with VARIOWIN should read this section before proceeding.

6.1 Data Files (.DAT)

Data files used by VARIOWIN programs are ASCII files conforming to the Geo-EAS [Englund & Sparks, 1991] or the GSLIB [Deutsch & Journel, 1992] file format.

- Line 1 holds the title of the file.
- Line 2 holds the number of variables, Nvar.
- Line 3 to Line 3 + (Nvar − 1) hold the name of each variable which cannot exceed 10 characters.
- All the subsequent lines contain sample values with variables listed in the same order as that used for listing the variable names. Values can be separated by BLANKS or by TABS. A sample name must be enclosed in ' (single quote) and must be in the last position.

According to the Geo-EAS convention, all values greater than or equal to 1.0E+31 are considered as missing values.

Example

Example.dat − Example file
5
X
Y
Arsenic
Cadmium
Lead
288.0 311.0 .850 11.5 18.25   'Sample 1'

285.6	288.0	.630	8.50	1.0e+32	'Sample 2'
273.6	269.0	1.02	7.00	20.00	'Sample 3'
...					
465.6	216.0	.930	11.6	25.00	'Sample 58'
492.0	216.0	.750	6.90	33.00	'Sample 59'
345.6	216.0	1.45	9.90	40.75	'Sample 60'

The end of the data file is indicated by the end-of-file character (EOF), which should be on the last sample line. However, Prevar2D and Vario2D with PCF will read data files having empty lines (i.e., lines not containing a digit) at the end of the file.

6.2 Pair Comparison Files (.PCF)

Prevar2D produces PCFs in a binary format that contains the following information writen in the order shown here:
- L (integer): length of the data file name, including a terminating NULL character
- Name (L bytes): data file name (*without* the directory path)
- Xcol (integer): x-coordinate column
- Ycol (integer): y-coordinate column
- F1 (integer): a flag (0 or 1) telling whether all variables are considered with their default limits. If a subset has been constructed by changing the default minimum or maximum value for one variable, this flag is set to 1.
- If F1 is set to 1:
  - Nvar (integer): number of variables in data file
  - For all variables ($i = 1$ to Nvar):
    - F2 (integer): a flag (0 or 1) telling whether the variable examined is considered with its default limits. If a subset has been constructed by changing the default minimum or maximum value for the variable, this flag is set to 1.
    - If F2 is set to 1:
      - min(V_i) (float): minimum value for variable i
      - max(V_i) (float): maximum value for variable i
- Npairs (long): number of pairs in PCF
- For all Npairs, which are ordered by increasing distances:
  - record(x) (integer): position in data file of the tail record
  - record($x + h$) (integer): position in data file of the head record
  - |h| (float): pair's distance
  - h_x (float): difference in x-coordinates
  - h_y (float): difference in y-coordinates

6.3 Grid Files (.GRD)

Grid files used for storing experimental variogram surfaces and 2D models of spatial continuity are ASCII files similar to the SURFER .GRD file format [Golden Software, Inc., 1994]. This type of file is used to transfer a grid to other software, such as Grid Display, which can display grids of values as pixel maps or as contour maps. A grid file contains the following information, written on different lines:

- id   id (4 characters) DSAA = ASCII grid file
- nx ny   nx (integer) = number columns (x-axis)
  ny(integer) = number of rows (y-axis)
- xlo xhi   xlo(double) = minimum x-coordinate of grid
  xhi(double) = maximum x-coordinate of grid
- ylo yhi   ylo(double) = minimum y-coordinate of grid
  yhi(double) = maximum y-coordinate of grid
- zlo zhi   zlo(double) = minimum z-coordinate of grid
  zhi(double) = maximum z-coordinate of grid
- grid row 1
- grid row 2
- grid row 3... (float): z-values of the grid organized in row order. Each row has a constant y-coordinate, with the first row equal to ylo, and the last row yhi. x-coordinates within each row range from xlo to xhi.

According to the convention used in SURFER, missing values are coded as 1.70141e+038.

Example

```
DSAA
11  11
–50 50
–50 50
31547.173828 138493.890625 95838.027389 110068.196685
111765.255632 107205.041694 94357.865728 85273.171618
81834.319259 86905.37505 81668.940854 88832.468786 82741.043904
                         ...
82741.043904 88832.468786 81668.940854 86905.37505 81834.319259
85273.171618 94357.865728 107205.041694 111765.255632
110068.196685 95838.027389
```

6.4 Variogram Surface Files (.VS)

Variogram surface files are ASCII files containing numerical results originating from the calculation of one or several (cross) variogram surfaces [4.4]. Several (cross) variogram surfaces can be written to the same file using the **Append** button of the dialog box appearing after an existing output file has been chosen under the **File | Save as...** menu item.

The grid corresponding to a (cross) variogram surface is written in the following manner:

- Line 1 contains a TAB and a string that indicates whether the grid refers to a variogram surface or to a cross variogram surface.
- Line 2 contains the number of lags along the x- and y-coordinates.
- Line 3 contains the lag spacing along the x- and y-coordinates.
- Line 4 contains the data covariance and the data correlation for cross variogram surfaces.
- Line 5 contains headers, separated by TABS, of information available for each grid cell:
  - DeltaX (double): x-coordinate of the cell
  - DeltaY (double): y-coordinate of the cell
  - NPairs (integer): number of pairs found for this cell
  - Mean |h| (double): mean pair distance for the cell
  - Variogram (double): (cross) variogram value [4.7.1]
  - Std. Variogram (double): standardized (cross) variogram value [4.7.2]
  - Covariance (double): (cross) covariance value [4.7.3]
  - Correlogram (double): (cross) correlogram value [4.7.4]
  - Madogram (double): (cross) madogram value [4.7.5]

Information relative to variogram surfaces only:
- mean(–**h**) (double): mean value of the variable calculated with the data found at the origin (tail) of the separation vector
- mean(+**h**) (double): mean value of the variable calculated with the data found at the end (head) of the separation vector
- var(–**h**) (double): variance of the variable calculated with the data found at the origin (tail) of the separation vector
- var(+**h**) (double): variance of the variable calculated with the data found at the end (head) of the separation vector

Information relative to cross variogram surfaces only:
- Z_1_mean(–**h**) (double): mean value of the first variable calculated with the data found at the origin (tail) of the separation vector
- Z_1_mean(+**h**) (double): mean value of the first variable calculated with the data found at the end (head) of the separation vector

- Z_2_mean(−**h**) (double): mean value of the second variable calculated with the data found at the origin (tail) of the separation vector
- Z_2_mean(+**h**) (double): mean value of the second variable calculated with the data found at the end (head) of the separation vector
- Z_1_var(−**h**) (double): variance of the first variable calculated with the data found at the origin (tail) of the separation vector
- Z_1_var(+**h**) (double): variance of the first variable calculated with the data found at the end (head) of the separation vector
- Z_2_var(−**h**) (double): variance of the second variable calculated with the data found at the origin (tail) of the separation vector
- Z_2_var(+**h**) (double): variance of the second variable calculated with the data found at the end (head) of the separation vector

All subsequent lines contain cell information separated by TABS. The grid is organized in row order with the first line corresponding to the row with the smallest DeltaY value.

All information, including that in the header lines, is separated by TABS.

Example

VARIOGRAM SURFACE
Variable: Z
Number of lags along X: 5   Number of lags along Y: 5
Lag spacing along X: 10   Lag spacing along Y: 10
Data variance: 8.99122e+04

DeltaX	DeltaY	NPairs	Mean l**h**l	Variogram	Std. Variogram
			Covariance	Correlogram	Madogram
			mean(−**h**)	mean(+**h**)	var(−**h**)
			var(+**h**)		
−50	−50	189	7.03221e+01	9.58380e+04	1.07134e+00
			−4.86763e+03	−5.44134e−02	1.74220e+02
			3.89619e+02	4.36180e+02	9.86675e+04
			8.11054e+04		
			...		
50	50	189	7.03221e+01	9.58380e+04	1.07134e+00
			−4.86763e+03	−5.44134e−02	1.74220e+02
			4.36180e+02	3.89619e+02	8.11054e+04
			9.86675e+04		

CROSS VARIOGRAM SURFACE
First variable: Z_1 Second variable: Z_2
Number of lags along X: 5   Number of lags along Y: 5

Lag spacing along X: 10   Lag spacing along Y: 10
Data covariance: 1.17768e+05 – Data correlation: 0.512

DeltaX	DeltaY	NPairs	Mean |h|	Variogram	Std. Variogram
			Covariance	Correlogram	Madogram
			Z_1_mean(−h)	Z_1_mean(+h)	Z_2_mean(−h)
			Z_2_mean(+h)	Z_1_var(−h)	Z_1_var(+h)
			Z_2_var(−h)	Z_2_var(+h)	
−50	−50	54	6.99072e+01	1.03949e+05	7.66160e−01
			2.34439e+04	1.97888e−01	1.96224e+02
			5.94576e+02	4.95122e+02	7.48343e+04
			6.04783e+04	6.68013e+02	3.19909e+02
			3.99203e+05	1.87550e+05	
			...		
50	50	54	6.99072e+01	1.03949e+05	7.66160e−01
			−2.72747e+04	−1.75535e−01	1.96224e+02
			4.95122e+02	5.94576e+02	6.04783e+04
			7.48343e+04	3.19909e+02	6.68013e+02
			1.87550e+05	3.99203e+05	

6.5 Variogram Files (.VAR)

Variogram files are ASCII files containing the numerical results from the calculation of one or several directional (cross) variograms [4.5]. Several directional (cross) variograms can be written to the same file using the **Append** button of the dialog box that appears after an existing output file has been chosen under the **File | Save as...** menu item. A directional variogram file containing several directional (cross) variograms is usually used to build a 2D general model of spatial continuity.

The table corresponding to a directional (cross) variogram is written in the following manner:

- Line 1 contains a TAB followed by a string that indicates whether the table refers to a directional variogram or to a directional cross variogram.
- Line 2 contains the name of the variable(s).
- Line 3 contains the directional parameters used to construct the directional (cross) variogram: direction, angular tolerance, maximum bandwidth.
- Line 4 contains the data covariance, the data correlation for cross variograms, and the code used for missing values.
- Line 5 contains headers, separated by TABS, of information available for each lag:
  - Lag (integer): lag number; 0 refers to the lag comprised between 0 and the lag tolerance

- NPairs (integer): number of pairs
- Mean |h| (double): mean pair distance
- Variogram (double): (cross) variogram value [4.7.1]
- Std. Variogram (double): standardized (cross) variogram value [4.7.2]
- Covariance (double): (cross) covariance value [4.7.3]
- Correlogram (double): (cross) correlogram value [4.7.4]
- Madogram (double): (cross) madogram value [4.7.5]

Information relative to directional variograms only:
- mean(–**h**) (double): mean value of the variable calculated with data found at the origin (tail) of the separation vector
- mean(+**h**) (double): mean value of the variable calculated with data found at the end (head) of the separation vector
- var(–**h**) (double): variance of the variable calculated with data found at the origin (tail) of the separation vector
- var(+**h**) (double): variance of the variable calculated with data found at the end (head) of the separation vector

Information relative to directional cross variograms only:
- Z_1_mean(–**h**) (double): mean value of the first variable calculated with data found at the origin (tail) of the separation vector
- Z_1_mean(+**h**) (double): mean value of the first variable calculated with data found at the end (head) of the separation vector
- Z_2_mean(–**h**) (double): mean value of the second variable calculated with data found at the origin (tail) of the separation vector
- Z_2_mean(+**h**) (double): mean value of the second variable calculated with data found at the end (head) of the separation vector
- Z_1_var(–**h**) (double): variance of the first variable calculated with data found at the origin (tail) of the separation vector
- Z_1_var(+**h**) (double): variance of the first variable calculated with data found at the end (head) of the separation vector
- Z_2_var(–**h**) (double): variance of the second variable calculated with data found at the origin (tail) of the separation vector
- Z_2_var(+**h**) (double): variance of the second variable calculated with data found at the end (head) of the separation vector

All subsequent lines contain lag information separated by TABS.
All information, including that in the header lines, is separated by TABS.

Example

VARIOGRAPHY

Variable: *V*

Direction: 14   Angular tolerance: 30   Maximum BW: 50

Data variance: 8.99122e+04   Code for missing values: –9999

Lag	NPairs	Mean |h|	Variogram	Std. Variogram	Covariance
		Correlogram	Madogram	mean(–**h**)	mean(+**h**)
		var(–**h**)	var(+**h**)		
0	66	3.65856e+00	3.43445e+04	5.72883e–01	2.89759e+04
		4.83331e–01	1.04063e+02	6.47048e+02	6.87248e+02
		8.02255e+04	4.47993e+04		
		...			
10	1564	1.00343e+02	8.62925e+04	1.18659e+00	4.56176e+03
		6.27279e–02	1.65871e+02	4.80610e+02	3.16925e+02
		1.04124e+05	5.07916e+04		

CROSS VARIOGRAPHY

First variable: Z_1 Second variable: Z_2

Direction: 14   Angular tolerance: 30   Maximum BW: 50

Data covariance: 1.17768e+05 Data correlation: 0.512   Code for missing values: –9999

Lag	NPairs	Mean |h|	Variogram	Std. Variogram	Covariance
		Correlogram	Madogram	$Z_1\_$mean(–**h**)	$Z_1\_$mean(+**h**)
		$Z_2\_$mean(–**h**)	$Z_2\_$mean(+**h**)	$Z_1\_$var(–**h**)	$Z_1\_$var(+**h**)
		$Z_2\_$var(–**h**)	Z_2 var(+**h**)		
0	49	3.58366e+00	8.10978e+04	3.75873e–01	7.75609e+03
		2.89447e–02	1.73835e+02	6.67396e+02	6.89110e+02
		7.43888e+02	7.68939e+02	8.39491e+04	4.85837e+04
		6.21206e+05	8.55326e+05		
		...			
10	437	1.00267e+02	1.34568e+05	7.35265e–01	1.04619e+04
		5.31340e–02	2.15955e+02	6.18927e+02	4.19644e+02
		8.08356e+02	5.11894e+02	1.03096e+05	3.73028e+04
		7.75840e+05	3.76044e+05		

6.6 Variogram Cloud Files (.CLD)

Variogram cloud files are ASCII files containing pairs displayed on a (cross) variogram cloud [4.6]. Several (cross) variogram clouds can be written to the same file using the **Append** button of the dialog box that appears after an existing output file has been chosen under the **File | Save as...** menu item.

The table corresponding to a (cross) variogram cloud is written in the following manner:

- Line 1 contains a TAB followed by a string that indicates whether the table refers to a variogram cloud or to a cross variogram cloud.
- Line 2 contains the name of the variable(s).
- Line 3 contains the minimum and maximum distances used in the calculation of the (cross) variogram cloud.
- Line 4 contains the directional parameters used to construct the (cross) variogram cloud: direction, angular tolerance, maximum bandwidth.
- Line 5 indicates the number of pairs found with the parameters indicated and the number of pairs that were displayed on the plot. Only the pairs displayed on the plot are written to the file.
- Line 6 contains headers, separated by TABS, of information available for each pair:
  - Rec(–**h**) (integer): record number in data file corresponding to the origin (tail) of the oriented pair
  - Rec(+**h**) (integer): record number in data file corresponding to the end (head) of the oriented pair
  - DeltaX (float): x-component of the separation vector
  - DeltaY (float): y-component of the separation vector
  - |**h**| (float): magnitude of the separation vector

  Information relative to variogram clouds only:
  - <variable>(–**h**): variable value at the origin (tail) of the oriented pair
  - <variable>(+**h**): variable value at the end (head) of the oriented pair
  - Variogram: variogram value of the pair [4.7.1]

  Information relative to cross variogram cloud only:
  - <first variable>(–**h**): first variable value at the origin (tail) of the oriented pair
  - <first variable>(+**h**): first variable value at the end (head) of the oriented pair
  - <second variable>(–**h**): second variable value at the origin (tail) of the oriented pair
  - <second variable>(+**h**): second variable value at the end (head) of the oriented pair
  - Cross variogram: cross variogram value of the pair [4.7.1]

All subsequent lines contain pair information separated by TABS.
All information, including that in the header lines, is separated by TABS.

Example

VARIOGRAM CLOUD

Variable: Z

Minimum distance: 0   Maximum distance: 100

Direction: 14   Angular tolerance: 30   Maximum BW: 50

10618 pairs in cloud – 10618 pairs on plot

Rec(–h)	Rec(+h)	DeltaX	DeltaY	|h|
		Z(–h)	Z(+h)	Variogram
439	266	–2.00000e+00	0.00000e+00	2.00000e+00
		5.97500e+02	7.42000e+02	1.04401e+04
		...		
32	107	1.00000e+02	0.00000e+00	1.00000e+02
		9.64000e+01	3.26000e+02	2.63581e+04

CROSS VARIOGRAM CLOUD

First variable: Z_1 Second variable: Z_2

Minimum distance: 0   Maximum distance: 100

Direction: 14   Angular tolerance: 30   Maximum BW: 50

3685 pairs in cloud – 3685 pairs on plot

Rec(–h)	Rec(+h)	DeltaX	DeltaY	|h|
		Z_1 (–h)	Z_1 (+h)	Z_2 (–h)
		Z_2 (+h)	Cross variogram	
439	266	–2.00000e+00	0.00000e+00	2.00000e+00
		5.97500e+02	7.42000e+02	9.97300e+02
		3.37150e+03	1.71536e+05	
		...		
264	311	8.00000e+01	6.00000e+01	1.00000e+02
		4.41000e+01	5.17300e+02	0.00000e+00
		1.48080e+03	3.50357e+05	

6.7 Model Files (.MOD)

Model files are ASCII files containing the model specifications displayed on a modeling window [5.3.1] as well as the parameters with which the directional (cross) variograms displayed on the plot window [5.3.4] were calculated. Model files can be created only when the modeling window is the active window. Several model specifications can be written to the same file using the **Append** button of the dialog box that appears after an existing output file has been chosen under the **File | Save as...** menu item. Information relative to a model is organized in the following manner:

- Line 1 contains the header MODEL PARAMETERS.

- Line 2 contains the name of the experimental measure of spatial continuity used to fit the 2D nested model [4.7].
- Line 3 contains the name(s) of the variable(s) for which the model was established.
- Line 4 contains the header EXPERIMENTAL VARIOGRAMS.
- The N following lines contain the parameters with which the N experimental (cross) variograms used for fitting the 2D model were constructed [4.5].
- The following two lines contain the description of the nested model that refers to the reduced vector and the anisotropy parameters [5.3.2].
- The next two lines contain the value of the indicative goodness of fit [5.3.3], the value of the a priori (co)variance used for the model, and the value of the a priori correlation.
- The next line contains the header MODEL PARAMETERS TO BE INCLUDED IN A GSLIB PARAMETER FILE.
- The subsequent lines describe the 2D general model in a GSLIB compatible format [Deutsch & Journel, 1992]. They can be cut and pasted into a parameter file used by a 2D kriging or simulation routine found in this geostatistical library.

Example

```
MODEL PARAMETERS
     Measure of spatial continuity: Variogram
     Variable(s): Z
EXPERIMENTAL VARIOGRAMS
     Direction: 14 – Tolerance: 30 – MaxBW.: N/A
     Direction: 104 – Tolerance: 30 – MaxBW.: N/A
     Direction: 59 – Tolerance: 30 – MaxBW.: N/A
     Direction: 149 – Tolerance: 30 – MaxBW.: N/A
Gamma (h): 22100 + 40300*Sph 25 (h) + 45500*Sph 50 (h)
          Dir.(1): 14 l anis.(1): 1.4 l Dir.(2): 14 l anis.(2): 2.9
IGF: 2.2977e–02
A priori covariance: 8.9912e+04          A priori correlation: 1e+00
MODEL PARAMETERS TO BE INCLUDED IN A GSLIB PARAMETER
FILE
2 22100    :nst, nugget
1 25 40300 76 1.4   :it, aa, cc, ang1, anis
1 50 45500 76 2.9   :it, aa, cc, ang1, anis


MODEL PARAMETERS
     Measure of spatial continuity: Variogram
     Variable(s): $Z_1 - Z_2$
EXPERIMENTAL VARIOGRAMS
```

Direction: 14 – Tolerance: 30 – MaxBW.: N/A
Direction: 104 – Tolerance: 30 – MaxBW.: N/A
Direction: 59 – Tolerance: 30 – MaxBW.: N/A
Direction: 149 – Tolerance: 30 – MaxBW.: N/A
Gamma (h): 46700 + 49300*Sph 25 (h) + 39100*Sph 50 (h)
              Dir.(1): 14 I anis.(1): 1 I Dir.(2): 14 I anis.(2): 1
IGF: 3.2222e–02
A priori covariance: 1.1776e+05   A priori correlation: 5.12e–01
MODEL PARAMETERS TO BE INCLUDED IN A GSLIB PARAMETER
FILE
2 46700   :nst, nugget
1 25 49300 76 1.1   :it, aa, cc, ang1, anis
1 50 39100 76 1.5   :it, aa, cc, ang1, anis

Further Reading

Deutsch, C. V. & Journel, A. G., "GSLIB Geostatistical Software Library and User's Guide," Oxford University Press, New York, NY, 1992.

Englund, E. & Sparks, A., "Geo-EAS 1.2.1 User's Guide," US-EPA Report #600/8-91/008, EPA-EMSL, Las Vegas, NV, 1991.

Golden Software Inc., "SURFER for Windows User's Guide," Golden Software Inc., Golden, CO, 1994.

Appendix A
Geostatistical Concepts

This appendix gives an overview of the geostatistical framework and introduces some key concepts without which the geostatistical approach cannot be properly understood and used. It also details at which stages of a geostatistical study VARIOWIN should be used. It has been included in order to remind the user of the necessary theoretical background required to understand the construction of a geostatistical model. Since this appendix serves only as a reminder, the reader is strongly advised to browse through the textbooks listed at the end of the appendix.

Within the geostatistical framework, a probabilistic model is used: the random function model. Estimations and/or simulations are performed using the properties of this model. The values or set of values produced by the estimation or simulation procedure are considered as a correct approximation of reality. Therefore, distinguishing between what belongs to the real world and what belongs to the probabilistic model is of paramount importance.

Figure A.1 displays the various steps involved in a geostatistical study. VARIOWIN provides tools to deal with the operations that appear in italic in the figure.

A.1 Random Variables, Regionalized Variables, and Random Functions

A *random variable* is a variable whose values are randomly generated according to some probabilistic mechanism. For instance, the result of casting a die can be considered as a random variable that can take one of six equally probable values.

A *regionalized variable* is a variable distributed in space; it is used to represent natural phenomena. For example, the heavy-metal content of the top layer of soil can be considered as a regionalized variable in the 2D space. A natural phenomenon, or a regionalized variable, usually shows a structured aspect as well as an erratic aspect:

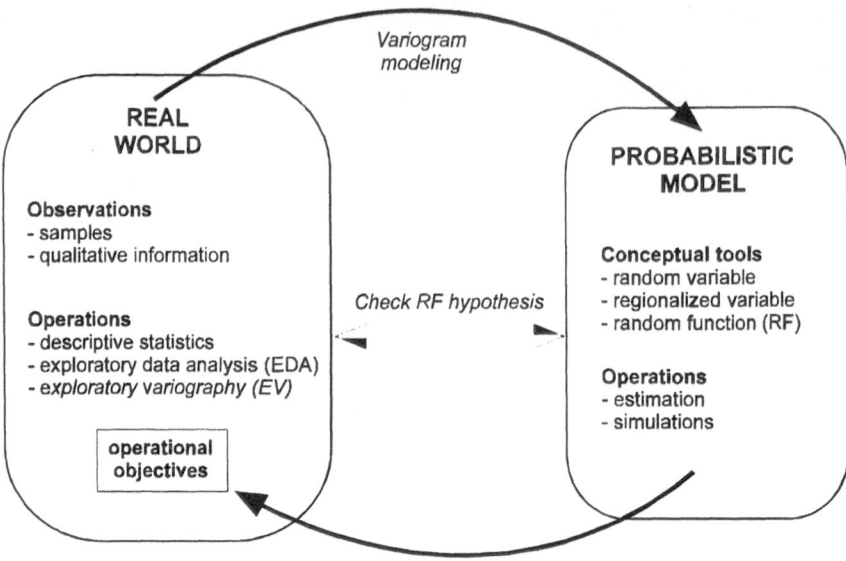

Figure A.1 Conceptual framework of the geostatistical approach.

- The *structured aspect* is related to the overall distribution of the natural phenomenon. For instance, in a polluted area, there are zones with, on average, a greater heavy-metal content than others.
- The *erratic aspect* is related to the local behavior of the natural phenomenon. For instance, within a given zone of a polluted area, the heavy-metal content seems to fluctuate randomly.

The formulation of a natural phenomenon must take this double aspect of randomness and structure into account. A consistent and operational formulation is the probabilistic representation provided by random functions.

A *random function* is a set of random variables $\{Z(\mathbf{x}) \mid$ location $\mathbf{x}$ belongs to the area investigated$\}$ whose dependence on each other is specified by some probabilistic mechanism. It expresses the random and structured aspects of a natural phenomenon in the following way:

- Locally, the point value $z(\mathbf{x})$ is considered as a random variable.
- This point value $z(\mathbf{x})$ is also a random function in the sense that for each pair of points $\mathbf{x}_i$ and $\mathbf{x}_i + \mathbf{h}$, the corresponding random variables $Z(\mathbf{x}_i)$ and $Z(\mathbf{x}_i + \mathbf{h})$ are not independent but are related by a correlation expressing the spatial structure of the phenomenon.

The regionalized variable representing the phenomenon (the set of all of its possible values distributed in space) is considered as a particular realization of the random function constructed on this phenomenon (the set of an infinity of random variables representing the values of the phenomenon at each point in space).

A.2 Moments Considered in Linear Geostatistics

Consider the random function $Z(\mathbf{x})$. For every set of k points in space $\mathbf{x}_1, \mathbf{x}_2, \ldots, \mathbf{x}_k$, there is a corresponding set of k random variables

$$\{Z(\mathbf{x}_1), Z(\mathbf{x}_2), \ldots, Z(\mathbf{x}_k)\}.$$

This set of random variables is completely characterized by the k-variable distribution function

$$F_{\mathbf{x}_1, \mathbf{x}_2, \ldots, \mathbf{x}_k}(z_1, z_2, \ldots, z_k) = \text{Prob} \{Z(\mathbf{x}_1) < z_1, \ldots, Z(\mathbf{x}_k) < z_k\}.$$

The set of all these distribution functions, for all positive integers k and for every possible choice of point $\mathbf{x}_i$, constitutes the *spatial law* of the random function $Z(\mathbf{x})$.

Exhaustive knowledge of the spatial law is not required for most estimation/simulation problems. Moreover, the amount of data is generally insufficient to infer the entire spatial law. In *linear geostatistics*, only the first two moments of the random function are used. Thus, no distinction will be made between two random functions having the same first- and second-order moments.

A.2.1 First-Order Moment – Mathematical Expectation

Consider a random variable $Z(\mathbf{x})$ at location $\mathbf{x}$. If the distribution function of $Z(\mathbf{x})$ has an expectation, then this expectation is generally a function of $\mathbf{x}$ and is written as

$$E\{Z(\mathbf{x})\} = m(\mathbf{x}).$$

A.2.2 Second-Order Moments

In geostatistics three second-order moments are considered:
1. the *variance* of $Z(\mathbf{x})$, which is the second order moment about the expectation $m(\mathbf{x})$ of the random variable $Z(\mathbf{x})$:

$$\mathrm{Var}\{Z(\mathbf{x})\} = \mathrm{E}\left\{[Z(\mathbf{x}) - m(\mathbf{x})]^2\right\}$$
$$= \mathrm{E}\left\{Z(\mathbf{x})^2\right\} - 2 \cdot \mathrm{E}\{Z(\mathbf{x})\} \cdot m(\mathbf{x}) + m^2(\mathbf{x})$$
$$= \mathrm{E}\left\{Z(\mathbf{x})^2\right\} - m^2(\mathbf{x}).$$

As with the expectation $m(\mathbf{x})$, the variance is generally a function of the location $\mathbf{x}$.

2. the *covariance*. If two random variables $Z(\mathbf{x}_1)$ and $Z(\mathbf{x}_2)$ have variances, then they also have a covariance, which is a function of the two locations $\mathbf{x}_1$ and $\mathbf{x}_2$:

$$C(\mathbf{x}_1, \mathbf{x}_2) = \mathrm{E}\left\{[Z(\mathbf{x}_1) - m(\mathbf{x}_1)] \cdot [Z(\mathbf{x}_2) - m(\mathbf{x}_2)]\right\}$$
$$= \mathrm{E}\{Z(\mathbf{x}_1)\} \cdot \mathrm{E}\{Z(\mathbf{x}_2)\} - \mathrm{E}\{Z(\mathbf{x}_1)\} \cdot m(\mathbf{x}_2) - \mathrm{E}\{Z(\mathbf{x}_2)\} \cdot m(\mathbf{x}_1)$$
$$+ m(\mathbf{x}_1) \cdot m(\mathbf{x}_2)$$
$$= \mathrm{E}\{Z(\mathbf{x}_1)\} \cdot \mathrm{E}\{Z(\mathbf{x}_2)\} - m(\mathbf{x}_1) \cdot m(\mathbf{x}_2).$$

3. the *variogram* or *semivariogram*. The variogram is defined as half the variance of the increment $[Z(\mathbf{x}_1) - Z(\mathbf{x}_2)]$:

$$\gamma(\mathbf{x}_1, \mathbf{x}_2) = \frac{1}{2} \mathrm{Var}\{Z(\mathbf{x}_1) - Z(\mathbf{x}_2)\},$$
$$2\gamma(\mathbf{x}_1, \mathbf{x}_2) = \mathrm{E}\left\{[Z(\mathbf{x}_1) - Z(\mathbf{x}_2)]^2\right\} - [\mathrm{E}\{Z(\mathbf{x}_1) - Z(\mathbf{x}_2)\}]^2,$$
$$\gamma(\mathbf{x}_1, \mathbf{x}_2) = \frac{1}{2} \mathrm{E}\left\{[Z(\mathbf{x}_1) - Z(\mathbf{x}_2)]^2\right\} \quad \text{when} \quad \mathrm{E}\{Z(\mathbf{x}_1) - Z(\mathbf{x}_2)\} = 0.$$

A.3 Ergodicity

The probabilistic framework deals with expected values, which are merely the integrals over all possible realizations of a random process. For an *ergodic* process, the expectation over all possible realizations is equal to the average over a single unbounded realization. Without the ergodic property, the moments of the random function could not be inferred from the spatial mean values calculated on the regionalized variable.

A.4 Hypothesis of Stationarity

In order to infer the spatial law or the first two moments of a random function $Z(\mathbf{x})$, where $\mathbf{x}$ is a point in space, many realizations $z_1(\mathbf{x})$, $z_2(\mathbf{x})$, . . ., $z_k(\mathbf{x})$, of $Z(\mathbf{x})$ would be required. However, unlike the casting of a die, it is not possible to obtain a new realization of the natural phenomenon at location $\mathbf{x}$. The reality is unique, and our model parameters must be inferred from this unique "realization".

If the phenomenon under study is homogeneous over a certain region, we can consider that its representation as a regionalized variable repeats itself in space. This repetition provides the equivalent of many realizations of the same random function $Z(\mathbf{x})$ and allows a certain amount of statistical inference. Two experimental values $z(\mathbf{x}_0)$ and $z(\mathbf{x}_0 + \mathbf{h})$ at two different points $\mathbf{x}_0$ and $\mathbf{x}_0 + \mathbf{h}$ can thus be considered as two different realizations of the same random function $Z(\mathbf{x})$ (Figure A.2).

Note also that such statistical inferences are possible only when the regionalized variable can be considered as ergodic.

This approach is not peculiar to geostatistics and random functions; it is more commonly used either to infer the probability distribution of a random variable from the histogram of data values or to infer the mathematical expectation from the average value of the data.

The hypotheses of stationarity are related to various degrees of the spatial homogeneity of the phenomenon under study. The type of stationarity that is assumed will tell what kind of statistical inference is permitted within the

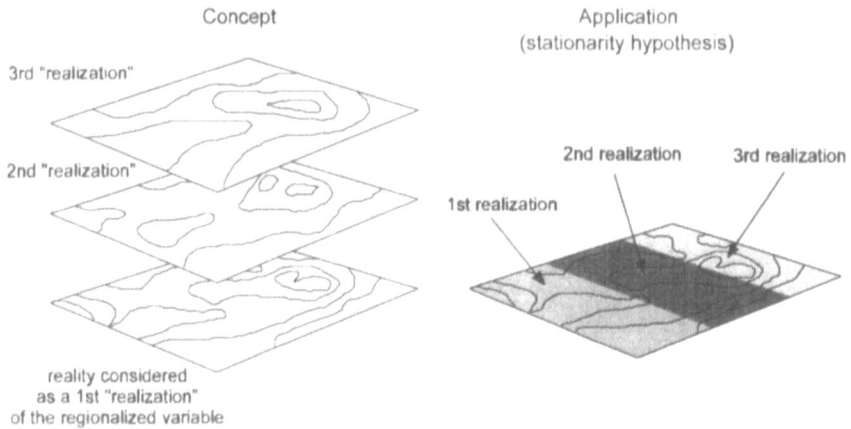

Figure A.2 Multiple realizations of a regionalized variable and the hypothesis of stationarity.

probabilistic model.

A.4.1 Strict Stationarity

A random function is said to be stationary, in the strict sense, when its spatial law is invariant under translation:

$$\{Z(x_1), Z(x_2), \ldots, Z(x_k)\} \quad \text{and} \quad \{Z(x_1 + h), Z(x_2 + h), \ldots, Z(x_k + h)\}$$

are identical in law (have the same k-variable distribution law) for any separation vector $\mathbf{h}$.

In practice, this very strong hypothesis is rarely assumed; it is usually replaced by the hypothesis of second-order stationarity or by the intrinsic hypothesis.

A.4.2 Second-Order Stationarity

Since linear geostatistics considers only the first two moments of the spatial law, it is sufficient first to assume that these moments exist and then to limit the stationarity assumption to them.

A random function $Z(\mathbf{x})$ is said to be second-order stationary if
- the mathematical expectation $E\{Z(\mathbf{x})\}$ exists and does not depend on location $\mathbf{x}$:

$$E\{Z(\mathbf{x})\} = m, \quad \forall \mathbf{x};$$

- for each pair of random variables $\{Z(\mathbf{x}), Z(\mathbf{x} + \mathbf{h})\}$, the covariance exists and depends only on the separation vector $\mathbf{h}$:

$$C(\mathbf{h}) = E\{Z(\mathbf{x}) \cdot Z(\mathbf{x} + \mathbf{h})\} - m^2, \quad \forall \mathbf{x};$$

The stationarity of the covariance implies the stationarity of the variance:

$$\text{Var}\{Z(\mathbf{x})\} = E\{[Z(\mathbf{x}) - m]^2\} = C(0), \quad \forall \mathbf{x}.$$

The stationarity of the covariance also implies the stationarity of the variogram:

$$2C(\mathbf{h}) = 2E\{Z(\mathbf{x}+\mathbf{h})\cdot Z(\mathbf{x})\} - 2m^2$$

$$= \left[E\{Z(\mathbf{x}+\mathbf{h})^2\} - m^2\right] + \left[E\{Z(\mathbf{x})^2\} - m^2\right]$$

$$- \left[E\{Z(\mathbf{x}+\mathbf{h})^2\} - 2E\{Z(\mathbf{x}+\mathbf{h})\cdot Z(\mathbf{x})\} + E\{Z(\mathbf{x})^2\}\right]$$

$$= 2C(0) - 2\gamma(\mathbf{h}),$$

$$C(\mathbf{h}) = C(0) - \gamma(\mathbf{h}).$$

This relation indicates that under the hypothesis of second-order stationarity, the covariance and the variogram are two equivalent tools for characterizing the correlation between two random variables $Z(\mathbf{x})$ and $Z(\mathbf{x}+\mathbf{h})$ separated by a vector $\mathbf{h}$.

The hypothesis of second-order stationarity assumes the existence of a covariance and, thus, of a finite a priori variance $Var\{Z(\mathbf{x})\} = C(0)$. Under this condition, the *correlogram*, $\rho(\mathbf{h})$, and the *standardized variogram*, $\gamma_S(\mathbf{h})$, can be defined and used similarly to the covariance and the variogram:

$$\rho(\mathbf{h}) = \frac{C(\mathbf{h})}{C(0)} = \rho(0) - \frac{\gamma(\mathbf{h})}{C(0)} = \rho(0) - \gamma_S(\mathbf{h}).$$

While those probabilistic functions are perfectly equivalent, their estimators are not. In practice, once the hypothesis of second-order stationarity has been accepted, one is free to use the probabilistic function that is the easiest to estimate in order to characterize the second moment of the random function $Z(\mathbf{x})$ [Srivastava & Parker, 1989].

Consequently, within the Model program, a 2D nested model of spatial continuity of a second-order stationary, random function can be fitted against any one of these measures of spatial continuity.

A.4.3 Intrinsic Hypothesis

Many physical phenomena have an infinite capacity for dispersion (i.e., brownian motion). These phenomena do not have an a priori variance or a covariance. However, the variance of their increments, or their variogram, exists and can be defined. As a consequence, the second-order stationarity hypothesis can be slightly reduced when assuming only the existence and stationarity of the variogram.

A random function $Z(\mathbf{x})$ is said to be intrinsic if

- the mathematical expectation $E\{Z(\mathbf{x})\}$ exists and does not depend on location $\mathbf{x}$,

$$E\{Z(\mathbf{x})\} = m, \quad \forall \mathbf{x};$$

- for all separation vector $\mathbf{h}$ the increment $[Z(\mathbf{x}+\mathbf{h})-Z(\mathbf{x})]$ has a finite variance which does not depend on $\mathbf{x}$:

$$\frac{1}{2}\mathrm{Var}\{Z(\mathbf{x}+\mathbf{h})-Z(\mathbf{x})\} = \frac{1}{2}E\{[Z(\mathbf{x}+\mathbf{h})-Z(\mathbf{x})]^2\} = \gamma(\mathbf{h}), \quad \forall \mathbf{x}.$$

The intrinsic hypothesis can also be seen as the limitation of the second-order stationarity to the increments of the Random Function $Z(\mathbf{x})$.

Note that second-order stationarity implies intrinsic behavior, but the converse is not true.

When the intrinsic hypothesis is accepted without the second-order stationarity hypothesis, only the variogram exists and can be used to characterize the second moment of the random function $Z(\mathbf{x})$.

Consequently, an experimental variogram that does not reach a plateau indicates that the underlying random function is not second order stationary. Within the Model program, a model of spatial continuity of such a random function may *only* be fitted against the variogram since the standardized variogram, the covariance, and the correlogram do not exist within the probabilistic framework (Figure A.1).

A.5 Why Do We Need to Model Variograms?

In linear geostatistics, values to be estimated or simulated are considered as a realization of a random variable that is a linear combination of known random variables:

$$Y = \sum_{i=1}^{n} \lambda_i Z(\mathbf{x}_i).$$

This linear combination of random variables is itself a random variable whose variance is given by

$$\mathrm{Var}\{Y\} = \sum_{i=1}^{n}\sum_{j=1}^{n} \lambda_i \lambda_j C(\mathbf{x}_i - \mathbf{x}_j).$$

The covariance function appearing in this expression must ensure that the variance will always be positive or equal to zero. By definition, such a covariance function $C(\mathbf{h})$ is said to be *positive-definite*.

Thus, not any function $g(\mathbf{h})$ can be considered as the covariance of a stationary random function: it must be positive-definite.

Assuming second-order stationarity [A.4.2], the variance of Y can be written in terms of the variogram:

$$\mathrm{Var}\{Y\} = C(0)\sum_{i=1}^{n}\lambda_i\sum_{j=1}^{n}\lambda_j - \sum_{i=1}^{n}\sum_{j=1}^{n}\lambda_i\lambda_j\gamma\left(\mathbf{x}_i - \mathbf{x}_j\right).$$

When the variance $C(0)$ does not exist, only the intrisic hypothesis [A.4.3] is assumed and the variance of Y is defined for *authorized linear combinations*, that is, linear combinations of random variables for which the sum of the weights is 0:

$$\sum_{i=1}^{n}\lambda_i = 0.$$

In this case, the $C(0)$ term is eliminated and the variance of Y is reduced to

$$\mathrm{Var}\{Y\} = \sum_{i=1}^{n}\sum_{j=1}^{n}\lambda_i\lambda_j\gamma\left(\mathbf{x}_i - \mathbf{x}_j\right).$$

The variogram function must be such that the variance of an authorized linear combination of random variables is positive or zero. By definition, $-\gamma(\mathbf{h})$ is said to be a *conditional positive-definite* function.

Using the property that $-\gamma(\mathbf{h})$ is a conditional positive-definite function, it can be shown that the semivariogram increases more slowly at infinity than the squared magnitude of the separation vector $\mathbf{h}$ does:

$$\lim\frac{\gamma(\mathbf{h})}{|\mathbf{h}|^2} = 0 \quad \text{when} \quad |\mathbf{h}| \rightarrow \infty.$$

As a consequence, an experimental variogram [4.5] whose variogram values [4.7.1] increase at least as rapidly as the squared magnitude of the separation vector is incompatible with the intrinsic hypothesis. This behavior most often indicates the presence of a *drift*, that is, a nonstationary mathematical expectation of the random function associated with the studied phenomenon.

This drift can often be subtracted from the data values using either a trend surface or a moving window map. The removal of the drift must be realized outside of VARIOWIN, which can then be used to perform the spatial data analysis of the residuals. All geostatistical estimations and simulations are then performed on the residuals, and the drift is added back afterwards.

A.6 The Multivariate Gaussian Random Function

This section describes one of the most often used geostatistical models as well as how VARIOWIN can be used to check that the properties attributed to the random function within this model are not in contradiction with the data.

The multivariate Gaussian model is extremely convenient because of its extreme analytical simplicity. It is also the limit distribution of many analytical theorems globally known as *central limit theorems*. If the spatial phenomenon is seen as being generated by the *addition* of several *independent* sources having similar spatial distributions, then the spatial distribution of the phenomenon can be modeled by a multivariate Gaussian random function:

$$Z(\mathbf{x}) = \sum_{k=1}^{K} Y_k(\mathbf{x});$$

where $Z(\mathbf{x})$ is the random function representing the phenomenon and $Y_k(\mathbf{x})$ are independent random functions having similar spatial distributions.

Natural processes that generate an observed phenomenon are rarely independent from one another and the way they interact is rarely additive. However, multivariate Gaussian models have been successful in a number of applications. This practical success combined with the extremely convenient analytical properties of the Gaussian models makes them the preferred choice for modeling continuous variables. Of course, if the multivariate Gaussian model is determined to be inappropriate, it should not be used.

The multivariate Gaussian model consists of the following two hypotheses:

- It is possible to transform the original variable into a new variable having a univariate standard normal distribution:

$$Z(\mathbf{x}) = \phi\{Y(\mathbf{x})\} \quad \text{or} \quad Y(\mathbf{x}) = \phi^{-1}\{Z(\mathbf{x})\},$$

where $Z(\mathbf{x})$ is the original random variable, $Y(\mathbf{x})$ is univariate standard normally distributed, and ϕ is a strictly increasing transformation function.

The transformation function, referred to as the *normal score transform* always exists. It is usually graphically defined by a one-to-one correspondence between the cumulative density function of the random

function $Z(\mathbf{x})$ and a standard normal cumulative density function [Verly, 1985].
- The multivariate distribution of the random function $Y(\mathbf{x})$ is Gaussian and strictly stationary [A.4.1].

As a consequence, the random function $Y(\mathbf{x})$ is fully determined by two parameters:
- its mean, $m(\mathbf{x})$:

$$E\{Y(\mathbf{x})\} = m(\mathbf{x}) = 0$$

for all locations $\mathbf{x}$;
- its covariance function, $C(\mathbf{h})$:

$$E\{Y(\mathbf{x}) \cdot Y(\mathbf{x} + \mathbf{h})\} = C(\mathbf{h})$$

for all locations $\mathbf{x}$;

Note that since $Y(\mathbf{x})$ is Gaussian, the hypothesis of strict stationarity is equivalent to the hypothesis of second order-stationarity [A.4.2].

The property that makes the multi-Gaussian random function such a pleasant model for geostatistical estimation and simulation is the fact that all conditional distributions of the random function $Y(\mathbf{x})$ are also multivariate normal. When dealing with spatial data, this means that the *local conditional distribution* at an unknown point (i.e., the cumulative density function at the unknown point, given the surrounding known values) can be reconstructed with only two parameters: its mean and its variance.

- The mean, or conditional expectation, is identical to the *simple kriging* (SK) estimate of that point value:

$$E\{Y(\mathbf{x}_0) \,|\, y(\mathbf{x}_\alpha) = y_\alpha, \quad \alpha = 1, \ldots, n\} \equiv \left[y(\mathbf{x}_0)\right]_{SK}^*$$

$$= m(\mathbf{x}_0) + \sum_{\alpha=1}^{n} \lambda_\alpha \left[y_\alpha - m(\mathbf{x}_\alpha)\right].$$

The n weights λ_α are given by the simple kriging system:

$$\sum_{\alpha=1}^{n} \lambda_\alpha C(\mathbf{x}_\alpha, \mathbf{x}_\beta) = C(\mathbf{x}_0, \mathbf{x}_\alpha), \quad \beta = 1, \ldots, n.$$

- The variance, or conditional variance, is equal to the simple kriging variance associated with the simple kriging estimate:

$$\text{Var}\left\{Y(\mathbf{x}_0) \,\big|\, y(\mathbf{x}_\alpha) = y_\alpha, \;\; \alpha = 1, \ldots, n\right\} = C(0) - \sum_{\alpha=1}^{n} \lambda_\alpha C(\mathbf{x}_0, \mathbf{x}_\alpha).$$

Note that this conditional variance does not depend on the data values but only on the data configuration and the covariance model.

The hypothesis of multinormality of the random function $Y(\mathbf{x})$ implies that each random variable [$Y(\mathbf{x})$, $\mathbf{x}$ being any location] is normally distributed. It also implies that each pair of random variables [$Y(\mathbf{u})$, $Y(\mathbf{v})$, $\mathbf{u}$ and $\mathbf{v}$ being any location] is binormally distributed, that each triplet of random variables [$Y(\mathbf{u})$, $Y(\mathbf{v})$, $Y(\mathbf{w})$, $\mathbf{u}$, $\mathbf{v}$, and $\mathbf{w}$ being any location] is trinormally distributed and so on.

Just as the strict stationarity hypothesis, the complete hypothesis cannot be tested. However, the multinormal distribution has many properties that can be used to design various checks:

1. Since any linear combination of a multinormal vector is also normally distributed, a check on binormality consists of verifying that the difference [$Y(\mathbf{x}) - Y(\mathbf{x} + \mathbf{h})$] is normally distributed with mean 0 and variance $2\gamma(\mathbf{h})$. Checking that all h-scatterplots [4.3] are elliptical clouds which show a greater density of points in the center and that are symmetric around the 45° bisector is a rough equivalent of this check of binormality.

2. For a multinormal random function, the conditional variance is independent of the conditional mean. As a consequence, multinormally distributed data should show no *proportional effect*, that is, there should not be a linear relationship between the local mean and the local variance. The presence of a proportional effect can easily be checked using *moving windows* statistics [Murray & Baker, 1991].

3. The sample indicator variogram can be compared with its theoretical bivariate counterpart (use the BIGAUSS program found in GSLIB [Deutsch & Journel, 1992]).

These three checks are quite effective. The first two are the easiest to realize. There is a fourth check, which is somewhat less effective, that uses the relationship between the experimental variogram values and the experimental Madogram values [4.7.5].

> Note that, in practice, only the bivariate normality of the random function $Y(\mathbf{x})$ is checked. Checking for trivariate, quadrivariate, K-variate normality would require many K-tuples sharing the same data configuration. This condition is rarely met in practice (unless the data are gridded and numerous), and the standard technique is to accept multivariate normality of the random function $Y(\mathbf{x})$ if its bivariate normality has been backed up by at least one of the four checks.

Further Reading

Chauvet, P., "Processing Data with a Spatial Support: Geostatistics and Its Methods," *Cahiers de Géostatistique*, Fasc. 4, Centre de Géostatistique, Fontainebleau, France, 1993.

Cressie, N. A. C., "Statistics for Spatial Data," John Wiley & Sons, New York, NY, 1991.

Cressie, N. A. C., "The Origins of Kriging," *Mathematical Geology*, Vol. 22, No. 3, pp. 239–252, 1990.

Deutsch, C. V. & Journel, A. G., "GSLIB Geostatistical Software Library and User's Guide," Oxford University Press, New York, NY, 1992.

Isaaks, E. H. & Srivastava, R. M., "An Introduction to Applied Geostatistics," Oxford University Press, New York, NY, 1989.

Journel, A. G., "Fundamentals of Geostatistics in Five Lessons," *Short Course in Geology*, Vol. 8, American Geophysical Union, Washington, DC, 1989.

Journel, A. G. & Huijbregts, C. J., "Mining Geostatistics," Academic Press, London, 1978.

Matheron, G., "Estimer et Choisir," *Les Cahiers du Centre de Morphologie Mathématique*, Fasc. 7, Ecole Nationale Supérieure des Mines de Paris, 1978.

Murray, M. R. & Baker, D. E., "MWINDOW: An Interactive FORTRAN-77 Program for Calculating Moving-Window Statistics," *Computers & Geosciences*, Vol. 17, No. 3, pp. 423–430, 1991.

Myers, D. E., "To Be or Not to Be . . . Stationary? That Is the Question," *Mathematical Geology*, Vol. 21, No 3, pp. 347–362, 1989.

Schofield, N., "Ore Reserve Estimation at the Enterprise Gold Mine, Pine Creek, Northern Territory, Australia. Part 2: The Multigaussian Kriging Model," *CIM Bulletin*, Vol. 81, No. 909, pp. 62–66, 1988.

Srivastava, R. M. & Parker, H. M., "Robust Measures of Spatial Continuity in Geostatistics," in *Geostatistics*, M. Armstrong (ed.), Proc. of the Third International Geostatistics Congress, September 5–9, 1988, Avignon, France, D. Reidel, Dordrecht, Holland, pp. 295–308, 1989.

Starks, T. H. & Fang, J. H., "The Effect of Drift on the Experimental Semivariogram." *Mathematical Geology*, Vol. 14, No. 4, pp. 309–319, 1982.

Strang, G., "Linear Algebra and Its Applications," Harcourt Brace Jovanovitch, San Diego, CA, 1976. Third edition, 1988.

Verly, G., "Estimation of Spatial Point and Block Distributions: The Multi-Gaussian Model," Ph.D. thesis, Stanford University, Stanford, CA, 1984.

Bibliography

Armstrong, M., "Common Problems Seen in Variograms," *Mathematical Geology*, Vol. 16, No. 3, pp. 305–313, 1984.

Barnes, R. J., "The Variogram Sill and the Sample Variance," *Mathematical Geology*, Vol. 23, No. 4, pp. 673–678, 1991.

Bradley, R. & Haslett, J., "Interactive Graphics for the Exploratory Analysis of Spatial Data – The Interactive Variogram Cloud," in Proc. of the Second CODATA Conference on Geomathematics and Geostatistics, P. A. Dowd & J. J. Royer (eds.), *Sci. de la Terre, Sér. Inf.*, Nancy, No. 31, pp. 373–386, 1992.

Carr, J. R., Bailey, R. E. & Deng, E. D., "Use of Indicator Variograms for an Enhanced Spatial Analysis," *Mathematical Geology*, Vol. 17, No. 8, pp. 797–811, 1985.

Chauvet, P., "Processing Data with a Spatial Support: Geostatistics and Its Methods," *Cahiers de Géostatistique*, Fasc. 4, Centre de Géostatistique, Fontainebleau, France, 1993.

Chauvet, P., "The Variogram Cloud," Internal Note N-725, Centre de Géostatistique, Fontainebleau, France, 1982.

Cressie, N. A. C., "Statistics for Spatial Data," John Wiley & Sons, New York, NY, 1991.

Cressie, N. A. C., "The Origins of Kriging," *Mathematical Geology*, Vol. 22, No. 3, pp. 239–252, 1990.

David, M., "Geostatistical Ore Reserve Estimation," Elsevier, New York, 1977.

Deutsch, C. V. & Journel, A. G., "GSLIB Geostatistical Software Library and User's Guide," Oxford University Press, New York, NY, 1992.

Englund, E. & Sparks, A., "Geo-EAS 1.2.1 User's Guide," US-EPA Report #600/8-91/008, EPA-EMSL, Las Vegas, NV, 1991.

Froidevaux, R., "Geostatistical Toolbox Primer, version 1.30," FSS International, Chemin de Drize 10, 1256 Troinex, Switzerland, 1990.

Golden Software Inc., "SURFER for Windows User's Guide," Golden Software Inc., Golden, CO, 1994.

Goulard, M. & Voltz, M., "Linear Coregionalization Model: Tools for Estimation and Choice of Cross-Variogram Matrix," *Mathematical Geology*, Vol. 24, No. 3, pp. 269–286, 1992.

Isaaks, E. H. & Srivastava, R. M., "An Introduction to Applied Geostatistics," Oxford University Press, New York, NY, 1989.

Isaaks, E. H. & Srivastava, R. M., "Spatial Continuity Measures for Probabilistic and Deterministic Geostatistics," *Mathematical Geology*, Vol. 20, No. 4, pp. 313–341, 1988.

Journel, A. G., "Fundamentals of Geostatistics in Five Lessons," *Short Course in Geology*, Vol. 8, American Geophysical Union, Washington, DC, 1989.

Journel, A. G. & Huijbregts, C. J., "Mining Geostatistics," Academic Press, London, 1978.

Matheron, G., "Estimer et Choisir," *Les Cahiers du Centre de Morphologie Mathématique*, Fasc. 7, Ecole Nationale Supérieure des Mines de Paris, 1978.

Microsoft Corporation, "Microsoft Windows 3.1 – Guide to Programming", Microsoft Press, Redmont, WA, 1992.

Murray, M. R. & Baker, D. E., "MWINDOW: An Interactive FORTRAN-77 Program for Calculating Moving-Window Statistics," *Computers & Geosciences*, Vol. 17, No. 3, pp. 423–430, 1991.

Myers, D. E., "Pseudo-Cross Variograms, Positive Definiteness and Cokriging," *Mathematical Geology*, Vol. 23, pp. 805–816, 1991.

Myers, D. E., "To Be or Not to Be . . . Stationary? That Is the Question," *Mathematical Geology*, Vol. 21, No 3, pp. 347–362, 1989.

Pannatier, Y., "VARIOWIN: Logiciel pour l'analyse spatiale de données en 2D – Etude géologique et géostatistique du gîte de phosphates de Taïba (Sénégal)," Ph.D. thesis, University of Lausanne, Lausanne, Switzerland, 1995.

Pannatier, Y., "MS-WINDOWS Programs for Exploratory Variography and Variogram Modeling in 2D," in *Statistics of Spatial Processes: Theory and Applications*, Capasso,V., Girone, G. & Posa, D. (eds.), Istituto per Ricerche de Mathematica Applicata (IRMA), Bari, Italy, pp. 165–170, 1994.

Posa, D., "Conditioning of the Stationary Kriging Matrices for Some Well-Known Covariance Models," *Mathematical Geology*, Vol. 21, No. 7, pp. 755–765, 1988.

Press, W. H., Teukolsky, S. A., Vetterling, W. T. & Flannery, B. P., "Numerical Recipes in C – The Art of Scientific Computing – Second Edition," Cambridge University Press, Cambridge, 1992.

Rendu, J., "Kriging for Ore Valuation and Mine Planning," *Engineering and Mining Journal*, Vol. 181, No. 1, pp. 114–120, 1980.

Schofield, N., "Ore Reserve Estimation at the Enterprise Gold Mine, Pine Creek, Northern Territory, Australia. Part 1: Structural and Variogram Analysis," *CIM Bulletin*, Vol. 81, No. 909, pp. 56–61, 1988.

Schofield, N., "Ore Reserve Estimation at the Enterprise Gold Mine, Pine Creek, Northern Territory, Australia. Part 2: The Multigaussian Kriging Model," *CIM Bulletin*, Vol. 81, No. 909, pp. 62–66, 1988.

Srivastava, R. M. & Parker, H. M., "Robust Measures of Spatial Continuity in Geostatistics," in *Geostatistics*, M. Armstrong (ed.), Proc. of the Third International Geostatistics Congress, September 5–9, 1988, Avignon, France, D. Reidel, Dordrecht, Holland, pp. 295–308, 1989.

Starks, T. H. & Fang, J. H., "The Effect of Drift on the Experimental Semivariogram." *Mathematical Geology*, Vol. 14, No. 4, pp. 309–319, 1982.

Strang, G., "Linear Algebra and Its Applications," Harcourt Brace Jovanovitch, San Diego, CA, 1976. Third edition, 1988.

Verly, G., "Estimation of Spatial Point and Block Distributions: The Multi-Gaussian Model," Ph.D. thesis, Stanford University, Stanford, CA, 1984.

Index